WONDERS

OF

ORGANIC LIFE.

PHILADELPHIA:
AMERICAN SUNDAY-SCHOOL UNION,
NO. 146 CHESTNUT STREET.
LONDON:
RELIGIOUS TRACT SOCIETY.

NOTE.—The *American Sunday-school Union* have made an arrangement with the *London Religious Tract Society*, to publish, concurrently with them, such of their valuable works as are best suited to our circulation. In making the selection, reference will be had to the general utility of the volumes, and their sound moral tendency. They will occupy a distinct place on our catalogue, and will constitute a valuable addition to our stock of books for family and general reading.

As they will be, substantially, reprints of the London edition, the credit of their general character will belong to our English brethren, and not to us; and we may add, that the republication of them, under our joint imprint, involves us in no responsibility beyond that of a judicious selection. We cheerfully avail ourselves of this arrangement for giving wider influence and value to the labours of a sister institution so catholic in its character and so efficient in its operations as the *London Religious Tract Society*.

☞ The present volume is issued under the above arrangement.

CONTENTS.

WONDERS

OF

ORGANIC LIFE.

CHAPTER I.

THE VITAL PRINCIPLE—THE BLOOD.

IN the consideration of organic beings, we must ever keep before us the great fact, that life is as much life in the monad or animalcule, as in the whale, the elephant, or the rhinoceros—that mass does not add intensity to vitality—that duration of existence is but an infinitesimal portion of time, whether it be counted by minutes or days, or measured by the revolutions of centuries. The elephant, and the ephemera; the banian tree of three thousand years, the sturdy oak, or churchyard yew, and the tender little annual that blooms and withers away, exist all and each according to prefixed laws; and when their extinction takes place, the time is as if it had never been. Startle not; for, O man! thou in the midst of creation standest alone, the sole mighty exception. To thee time will not be as if it had never been; it is with thee the precursor of a momentous eternity.

It has been often said, wherever *life* can be there life is. The minutest drop of water is to some beings an ample sea; everywhere around us, on the land, in the water, in the air, the results of the creative fiat are made manifest; and the number and diversity of organic beings overwhelm us with astonishment. The huge whales of the ocean present us with the largest of animal forms, and the giant trees of the inter-tropics with the most stupendous of vegetable productions; but, on the other hand, it has not been determined at what degree of minuteness a boundary is put to organization. The most powerful microscopes are limited in their sphere, yet they have opened to us domains of organic life, the existence of which without their aid would never have been suspected. By the term life, as we here use the word, we mean organized beings in contradistinction to inorganized matter, even in a state of crystallization.

It may, however, be asked in the outset, What do we understand by *life* in the abstract, that is, irrespective of organization? We shall answer the question as best we can—craving our reader's indulgence, if, in the explanation of subjects of an abstract nature, we are occasionally necessitated to employ language of a scientific character.

Life, it must be admitted, prosecute our researches into it as we may, is a deep mystery. Setting *man* aside, we know not what *any* animal loses when it dies, nor what it had before it died; therefore we know not what death is, further than is made manifest to the senses of the living, and that is, a resolution of the frame into the elements around us—dust unto dust. But *something* once kept this dust together—gave it feelings, passions, and desires —rendered the assumption of nutritive particles imperative, and made reproduction a law—such was the *fiat* of creation. What, then, is this *something?*—this that makes the infant grow to manhood, the acorn rise into the oak—this that permits man to wither as the flower, and the oak to moulder into ruin. What, we repeat, is *life?* —or, if the reader like the term better—What is the vital principle? It is obviously either something or nothing; if something, it must be superadded to organization; if nothing, it must be a *consequence* of organization, and a mere aspect of matter under certain conditions or arrangements. But inert matter cannot vitalize itself, nor can any other than vitalized bodies produce or generate vitalized bodies. From vital organization alone, is vital organization transmitted. Turning for a solution of the difficulties which invest the subject to an examination of the phenomena of death, mystery is still found enshrouding it.

The precise mode in which an animal dies we cannot tell; to say that it has ceased to breathe and to feel, is to say nothing—for these things are merely the consequences of death. We may kill by blood-shedding, or by strangulation, or by the infliction of agony; but even then, the question reverts—What has quitted the body, leaving it a prey to the operative influence of the laws of chemical dissolution? The whole subject, we are forced again to confess, is shrouded in mystery. This we know, that vitality and organization (a convenient term, if understood in its true sense) are ever associated—that when vitality becomes extinct, organization retains an aspect only, waiting for the laws of chemistry to do their office—and that living beings only can produce living successors. One thing is certain, that between living bodies and the laws of chemical affinity there is, as far as vitality is concerned, a great antagonism, for organic resolution without commensurate reparation is the consequence only of death. Nevertheless, at the same time that we urge this grand principle, we acknowledge that chemical actions are perpetually in operation in all living bodies, but under the mystic laws of the vital energy. It must be at once apparent, then, that chemical operations or changes, however intricate, however precise, displayed by organic bodies, cannot be the original cause of *vitality*, nor yet be, in and of themselves, the vital principle. The same observation applies to that recondite agent which we term electricity, galvanic action, or electro-magnetism. No action in the organic living frame takes place without involving galvanic or electric changes. The nervous system is a wonderful galvanic apparatus, but *life* is here again a cause, not a consequence; the *truly dead* nerve, when vital warmth has left the body, ceases to respond to the experimental appliances of the physiological anatomist; the sheep, as is often to be seen, may, when just slaughtered, be made to start and struggle, and breathe forcibly, by means of the agency of electric currents skilfully thrown upon certain nerves. The frog, immediately killed, may be made to leap by similar means; but this galvanic irritability (*not sensibility*) of the nerves, and contractility of muscle, are merely the remains of a power implanted in them by some other principle, and are not inherent in the matter composing these organs; otherwise how could it be lost? This principle we call life, or vitality, terms expressive rather of our ignorance than of our knowledge.

We have shown in a few words what life *does*, rather than what life *is*. Let us see if Cuvier, one of the greatest physiological anatomists and philosophers of any age, has done more. "If, in order," he says, "to obtain a just idea of the *essence of life*, we will consider it in beings wherein its effects are the most simple, we shall quickly perceive that it consists in the faculty which certain organic combinations possess of existing, during a certain length of time, and under a determinate form, whilst attracting continually into their composition a portion of surrounding matter, and restoring to the elements portions of their own constituent body. Life is a revolving vortex, more or less rapid, more or less complicated, the direction of which is constant, and which ever draws in molecules of affinity; but at the same time into which individual molecules enter, and from which they are as continually thrown off, insomuch as that the form of a living body is more its essential, than is its constituent matter. As long as this movement subsists, the body in which it is in activity is *living —it lives;* as soon as the movement ceases for good, the body *dies.* After death the elements which compose it, rendered up to the ordinary affinities of chemistry, delay not to separate from each other, whence results more or less the dissolution of the body which has been once living. It was, then, by the vital movement that dissolution was restrained, and that the corporeal elements were for a space bound in union.

"All living bodies die after a space of time, the extreme limit of which is determined for every species; and death appears to be a necessary effect of life, which, by its own action even, essentially alters the structure of the body in which it operates, so as to render its continuance impossible." Yet again, therefore, the question reverts—What is life, and wherefore should the frame be worn out by the vital actions of its own machinery? Cuvier tells what the results of vitality are, as far as we can appreciate them, but not what causes the ever-changing combinations, the attractions and repulsions which are perpetually going on in organic bodies. Life—we are once more forced to the conclusion—is a mystery; nor can we penetrate beyond the revelation which God has made in the first chapter of the book of Genesis, wherein we learn only this, that living beings arose into existence in obedience to his word. In their Introduction to the French edition of Meckel's Comparative Anatomy, the translators (MM. Riester and Alph. Sanson) ask—"The principle of life, is it anything more than a modification of the cause of electric forces?" We do not quite understand the meaning of the latter phrase; however, the writers state that this is far from being definitely settled, and then they go on to observe, "that chemistry has revealed the primitive elements of organic bodies, and that microscopic observation has demonstrated the globular disposition as the essential form of the intimate structure of all tissues under the influence of life." We may here reiterate our observation, that although the vital phenomena may involve a perpetual play of galvanic or electro-galvanic changes and phases, and may agitate the nerves so as to render them efficient in the fulfilment of their multifarious offices, we have no reason thence to assume that the electric fluid under any aspect is identical with the vital principle.*

* "Wherever there is organization and life there is also electric tension, or the play of the voltaic pile, as the experiments of Nobili and Matteucci, and especially the latest admirable labours of Emil du Bois, teach us. The last-named philosopher has succeeded in manifesting the presence of the

We read in the Scriptures, "The life of the flesh is in the blood," Lev. xvii. 11; and further on, "The life of all flesh is the blood thereof," v. 14. By this, without straining the point, we may infer that it is in the circulating fluid, blood, sanies, or juice, as the case may be, of all organic bodies, that the living principle exists. Blood is a living fluid, destined to repair the solid body, to afford materials for the building up of every tissue, bone, or muscle, and to recruit itself in a mysterious manner, by the conversion of nutriment into its own character, and by the action of the atmosphere or the water. The very nerves themselves are derived from the contributions afforded by the blood. If the heart ceases to act, we die; if the blood be drained to a certain extent from the system, we die; if its waste be not duly repaired, we languish and sink; if the refreshment of the blood, by respiration, be prevented, we die suffocated. Stop the circulation of a plant, ring the bark of a tree with the knife, and every part above fades, withers, and perishes. Some animals of the lower orders

electric muscular current in living and wholly uninjured bodies. He shows that the human body, through the medium of a copper wire, can cause a magnetic needle at a distance to be deflected at pleasure, first in one, then in the opposite direction. I have witnessed these movements produced at pleasure, and have had the gratification of seeing thereby great and unexpected light thrown on phenomena, to which I had laboriously and hopefully devoted several years of my youth."—*Humboldt.*

Within the last few years this interesting subject has received great attention, and many curious facts have been elicited, of physiological importance, but into which we cannot here fully enter.

seem to consist of little else than a quantity of vital fluid. Of such are the large *Medusæ* or jelly fish, which we see floating on the surface of the sea. They are composed of a most delicate cellular tissue, filled with fluid; and this being exhaled, as it often is, when individuals are thrown upon the shore, and exposed to the fervid rays of the sun, all that is left is a slight filmy shred, a few grains only in weight.

It is evident that the vital principle, in an abstract sense, is beyond our research; we can only study it in its phenomena; we can only watch its effects and trace out its operations. The Scripture, as we have seen, declares that the "life of the flesh is in the blood." It will be interesting, therefore, without entering very abstrusely into the subject, to inquire as to the composition of this fluid, at least in the higher animals, and examine the mode in which its losses are recruited, and its refreshment effected; the more so, as in pursuing our inquiries it will become manifest that both chemical and galvanic changes or agencies are involved in the phenomena connected with its waste and renewal.

The blood of man or quadruped, when poured out from the arteries or veins, is a red viscid fluid, which, if suffered to remain for an hour or two at rest in a vessel, separates, on losing vitality, into two parts, one fluid, called *serum*, the other tolerably solid, called clot or *crassamentum*. Venous and arterial blood differ from each other in colour, the former being of a dark modena-red, the latter of a bright scarlet. The specific gravity of venous blood is greater than that of arterial, and this appears to be owing to the carbon contained in the latter. The specific gravity of the blood, averaging 1050, (water being 1000,) is capable of increase or decrease, according to health and diet; and in the higher animals it always exceeds that of the lower. Its temperature is greater in birds than in any other warm-blooded animals.

The separation of blood into serum and crassamentum is the result of its death, or perhaps the mode of its dying. During this process a peculiar odour is exhaled, in the form of a vapour, from which we turn away with aversion, and from which the ox instinctively recoils when forced into the slaughter-house. The quantum amount of serum in proportion to the crassamentum varies in different animals, and in the same animal, according to its condition and state of health. The serum is most abundant in small feeble animals, destitute of energy; small in quantity in animals of muscular vigour and ferocious habits. Its general colour is yellowish, often with a tinge of pale green; it is adhesive, and has a saline flavour. In its chemical composition it is essentially albumen, and coagulates by heat. At the temperature of 160° it is converted into a substance similar in appearance and character to the *white* of a hard-boiled egg. This fluid albumen contains several earthy and neutral salts in a state of solution, such as hydrochloride of soda and potass, subcarbonate and phosphate of soda, sulphate of potass, phosphate of lime, magnesia, and iron, with subcarbonate of lime and magnesia. It also yields an oily and a crystallizable fatty matter. The crassamentum, on its separation from the serum, assumes the form of two layers, not, however, rigidly distinct from each other, but rather blending together. The upper layer appears in the form of a yellowish white glaze, or tenacious skin, and on minute examination will be found to be moderately firm, tough, and elastic, and to consist of a vast number of minute threads or fibres, disposed in various directions, crossing and recrossing each other. From this circumstance it has received the expressive term of *fibrin.* Under microscopic analysis these fibres are precisely similar to those of a muscle when deprived of its enveloping membrane and its colouring matter. The fact is that the two are identical, for the basis of all muscles, and of the solid parts of the body generally, the bones excepted, consists of *fibrin.* It is the most important constituent of the blood; whatever other may be absent, this is invariably present in the blood of all animals which possess blood, whether coloured or colourless. Below the glaze of *fibrin,* which varies in thickness under influencing circumstances, we find a deep red mass, consisting of the colouring particles of the blood, which being of a greater specific gravity than *fibrin,* have gradually subsided during the progress of coagulation. When examined under a microscope of great power, this mass is found to be composed of extremely minute corpuscles, varying in size in different animals. These corpuscles are in the form of flattened discs, circular, oval, or elliptical, and surrounded by a most delicate envelope. According to Mr. Lister and Dr. Hodgkin, these red particles are solid flattened bodies, without any envelope. In the lower vertebrate animals, however, there is a distinct and permanent capsule surrounding a nucleus. In the human subject these blood-discs are circular, flattened, and rather concave on each side, with a rounded margin. Wollaston estimates them at the five-thousandth part of an inch in diameter; others, however, give different estimates of their admeasurement, and among them Hodgkin and Lister, who set down the size as three thousand. It would appear, however, that some variableness as to size in the blood-discs of the same clot is observable; hence, perhaps, arises the discordance in question. Within the last few years Mr. Gulliver has devoted great attention to this subject;* and it appears, from this gentleman's observations, that in the mammalia generally the blood-discs are circular, but in some oval. In birds, reptiles, and fishes, their figure is either oval

* We refer to the Med. Chir. Trans., vol. xxiii. Dublin Med. Press, Nov. 27, 1839. Annals of Nat. Hist., Dec. 1839. Lond. and Edin. Phil. Mag., 1839. Appendix to Gerber's Anatomy. The Proceeds. Zool. Soc. Lond. 1840 to 1846, etc.

or elliptical; and in fishes they are larger than in any of the other vertebrate classes.

It may be observed, that besides the red blood-corpuscles, pale globules present themselves under the microscope—globules of fibrin; and between these two there appears to be a sort of repulsion preventing their union, and thus maintaining the blood in a fluid state. But when drawn from the body, the blood soon coagulates, this self-repulsion ceasing, and cohesion between the corpuscles taking place. The red particles are composed of a substance resembling fibrin, to which the term *globuline* has been given; they contain a red colouring matter, called *hæmatosine*, and a certain variable quantity of oxide of iron. While from the fibrin of the blood the muscles and general tissues of the body are formed, it would seem that the red corpuscles act the part of carriers of oxygen from one part of the system to another, and are thereby the active agents by which animal heat is kept up; and it is observable, that the greater the energy and activity of the animal, the greater is the proportion of this red matter, and also the elevation of the animal temperature. It is well known that there are many animals of a soft or gelatinous consistence, in whose structure no muscles or distinct muscular fibres can be detected, although they are capable of executing decided movements. In such cases it is very probable that motion is produced by changes in the state of the particles of fibrin, in their temporary approximation to each other, or in some alteration which we cannot appreciate, but which enables them to perform the function of muscles. We have said that the coagulation of the blood is the result of its death, or the mode in which it parts with its vitality. But in the case of animals struck dead by lightning, the blood does not coagulate, the muscles do not become rigid, and decomposition rapidly ensues. Here the vitality of the blood is instantaneously destroyed; no space of time being allowed for it to exhibit the ordinary phenomena attending its gradual death. Of the changes which the blood undergoes in various diseases, or disturbances of the organic functions from unnumbered causes, it is not here our place to speak; our aim is solely to give an explanation of the composition of the vital fluid, of which many persons are ignorant. From the blood the system is nourished and repaired; perspiration and all the secretions tend to its decrease; hence it continually needs recruiting. This is effected by the reception and assimilation of matter which had once lived under an animal or vegetable form, and which having died is now to live again, but also again to die, and perchance enter into another condition of organization.

The process termed *digestion*, by which the blood is recruited, remains to be briefly detailed. All animals do not crush, mince, or masticate the food previously to transferring it to the stomach; reptiles and fishes swallow their prey whole, and the lower animals engulf their victims at once into their digestive cavity. Granivorous birds, also, swallow their food whole, but the carnivorous tear the flesh into strips or morsels, and swallow it. In the parrot, however, something like mastication is observable, the mobility of the mandibles of its beak enabling it thus to treat the food upon which it naturally subsists. The toucan, too, crushes and squeezes with its beak the unfortunate bird which it has seized before swallowing it, and this may be called a sort of mastication. Among mammalia, neither the dog or wolf, nor the lion or tiger, can be said to masticate their food; they rend the flesh into portions, which are bolted at once. The whales and grampuses, also, swallow their prey at once. On the other hand, in such animals as the horse, the ox, the sheep, the deer, and others, mastication is carefully performed, and their teeth are expressly fitted for a grinding action, the salivary glands, too, being large. Man, also, masticates, or ought to masticate his food; for this process, during which the food becomes of a due temperature, and is mixed with a certain portion of saliva, is requisite for its easy digestion.

It is to the digestive process, as carried on in the human species, that we shall here confine our attention. In all other animals the process is essentially the same, and involves the agency of a chemical solvent. The food crushed by the teeth, and mixed with saliva, is transferred to the stomach, where it is maintained at the temperature of 100° of Fahrenheit, and kept in a state of gentle but almost unceasing agitation, by a peculiar motion of that organ, effected by its muscular fibres. This vermicular motion is termed *peristaltic*. The stomach, we may here observe, is divided into two portions, a cardiac portion, into which the gullet immediately leads; and a pyloric portion, which opens into the commencement of the alimentary canal. Now, the food when swallowed passes into the cardiac portion; and the stomach, which when unemployed is one undivided bag, or sacculus, as it is termed, contracts after the fashion of an hour-glass, by the action of the circular fibres of the muscular coat, thus forming two bags, or sacculi, by far the largest of which is the cardiac. Here the food awaits the dissolving influence of the gastric juice. At this juncture a remarkable change takes place in the lining membrane of the stomach. When the stomach is empty, this membrane is of a pale pink colour; but now it becomes of a bright red colour, studded with innumerable minute lucid points, from which a pure limpid and colourless fluid distils, and mingling with the food effects its gradual solution. This fluid is the gastric juice, the true solvent of the food, and its action is entirely chemical; as the food is dissolved, the reduced material is transmitted by means of the peristaltic action of the muscular coat of the stomach into the pyloric portion, where it accumulates. This solution of the food is in no respect analogous to that decomposition by putrefaction which would be effected by the agency of warmth and moisture. It is a true chemical solution, and what is more, the gastric juice is highly antiseptic, and will immediately arrest the putrefactive process, even after it has advanced to a considerable extent. By a wise provision this solvent will not act upon the living stomach, but it will act upon the stomach after death; and many cases are on record in which the coats of the stomach have been more or less extensively eaten away, and the adjacent parts rendered soft and pulpy by its agency, a fact well known to the celebrated John Hunter.

With regard to the nature of this fluid, there is now only one opinion—it is acid; this was clearly ascertained by Spallanzani. Dr. Prout was, we believe, the first to prove that this acid is the muriatic; his experiments have been confirmed by those of Tiedemann and Gmelin, and more recently by those of Braconnot and Blondelot; so that it may now be regarded as an established fact, that muriatic acid, or chlorine, (in a diluted state,) is the agent by which the solution of the food is effected. It is a remarkable fact, that if meat be enclosed in a glass tube with diluted muriatic acid, and kept at the temperature of the blood, it will be converted into a uniform semi-fluid mass, closely resembling that formed by the action of the gastric juice on food in the stomach, and which is termed *chyme*.

The muriatic acid or chlorine, which is thus proved to be the essential ingredient of the gastric juice, is with reason supposed to be derived by the secreting agency of minute glands from the common salt (muriate of soda, or more correctly, hydrochloride of sodium) contained in the blood. If this be the case, the salt must undergo a complete decomposition, and its metallic base, *sodium*, must enter into some new combination. The chyme into which the gastric juice has reduced the food is an acid semi-fluid, or pultaceous mass, usually of a greyish colour; it exhibits, however, some differences according to the nature of the food, both as to colour and consistence, but it is invariably acid. In this state it is gradually transmitted from the pylorus, through the pyloric orifice, into the commencing portion of the alimentary canal, termed the duodenum. Here it undergoes further changes. It becomes mixed with the mucous secretion of the canal, with the pancreatic juice, a fluid somewhat resembling saliva, and with bile poured drop by drop from a fine duct, which thus conveys the secretion of the liver to the duodenum. A singular transformation now begins to take place in the *chyme;* it gradually separates into two portions, namely, a whitish tenacious fluid, sometimes opaque, and of an alkaline quality, termed *chyle,* and a pultaceous residuary portion. The chyle is that portion of the food which is destined to become blood; the remainder is useless.

By means of the peristaltic action of the small intestines, both the chyle and the residuary matter are gradually carried forwards; the progress of the former is peculiarly slow, for it adheres from its tenacity to the villi (or velvet-like pile) of the inner coat of the intestine, and is, moreover, obstructed by certain valvular folds, termed *valvulæ conniventes.* All this is by design, as it affords time for a system of vessels, termed lacteals, to absorb it. These lacteals, so called from the milky nature of the fluid they contain, commence by innumerable open mouths on the surface of the villi; they then pass through the coats of the intestine, and run through the layers of the mesentery, enter into the first series of mesenteric glands, in which they become extremely convoluted, and communicate freely with each other. After emerging from these glands, the chyliferous vessels continue their course between the layers of the mesentery, and enter a second series of glands, in which they again become convoluted. On freeing themselves from the second set of glands, they converge into a receptacle for the chyle, which constitutes the commencement of the thoracic duct. Here we must observe, that in this receptacle also terminate another set of vessels, termed absorbents, and also lymphatics, (from the colourless and pellucid fluid, or lymph, which they contain.) These absorbents arise from every portion of the frame, and are ever active in removing the materials of which it is composed, the minute arteries depositing fresh materials in their place, so that the body is in a perpetual process of change, by agencies within itself. The absorbents or lymphatics, as we have said, enter the receptacle for the chyle, and the lymph and chyle become mingled together. The thoracic duct, in its course upwards, receives other lymphatic vessels, the contents of which also mingle with the chyle, and are poured into the venous blood near the heart. This commixed fluid immediately passes into the right cavities of the heart, and is thence sent to travel through the minute vessels which ramify on the cellular tissue of the lungs. Here the chyle undergoes its last conversion—it becomes living arterial blood. What changes the chyle undergoes in the mesenteric glands is not understood, nor is the use of the pancreatic juice satisfactorily ascertained. According to Dr. Prout, this fluid contains albumen, and a curdy substance; it is slightly acid, and holds in solution matters of a saline nature.

If the reader asks how the process which we have briefly detailed can turn particles of aliment, dead matter, into living blood, we can only answer, that we do not know; we *know* that such is the process; the rest is shrouded in mystery.

Here, perhaps, as a sequel to the general details which we have attempted to sketch, is the most fitting place for some observations respecting a principle in the composition of animal and vegetable organization, and which exists alike in albumen, casein, (cheese,) horn, animal fibrin, and vegetable fibrin, discovered in 1838 by Mulder, the Dutch chemist, who gave to it the term of protein, (from the Greek πρωτευω, prōteuō, I take the first place,) because he regarded it as the basis of all the other substances, or living tissues. In the same year professor Schleiden, of Jena, "gave the results of his researches on the formation of the cells in plants, and pointed out the existence of a nucleus, or cell bud, from which each cell originated. This nucleus, according to Schleiden, is always composed of protein; so that it would appear that the protein is the earliest formed of vegetable substances, and is the seat of that residual power which, in the absence of any intimate knowledge of its nature, is called *vitality*." Protein is composed of carbon forty-eight parts, hydrogen thirty-six, nitrogen, or azote, six, and oxygen fourteen; but there appears to be some diversity of opinion as to the exact proportions of each of these primary elements; at the same time, whether this substance be obtained from animal or vegetable matters, analysis proves that very little difference of elementary composition is in either case to be detected. When either albumen or casein, horn, or animal or vegetable fibrin, is dissolved in a solution of potass, and this solution, after filtration, is mixed with a slight excess of acid, a greyish white flocculent precipitate is abundantly formed, and a slight smell of hydrosulphuric acid is perceived. This flocculent substance is protein. It exhibits the following properties:—The white flocculi, while moist, are semi-transparent, but on being dried they become yellowish, hard, and brittle. Protein is inodorous and tasteless ; it absorbs moisture rapidly from the air, and loses water at 212°. It is insoluble in water, ether, alcohol, or in essential oils. Nevertheless, by long-continued boiling in water, it undergoes some change of properties, and is rendered soluble. Protein is, however, dissolved by acetic and phosphoric acids, and also by hydrochloric acid, the solution in this case having a tint of indigo, which changes to black when subjected to heat. Under the action of concentrated sulphuric acid it produces a jelly, which contracts in water ; and this jelly, after being washed in water and alcohol, retains a minute portion of acid, though not sufficient to redden litmus-paper. Mulder calls this compound *sulpho-proteic acid.* When protein is boiled in dilute sulphuric acid, it acquires a purple tint.

Protein, then, is the essential nutriment which animals derive from plants or grain, or from other animals. It constitutes, moreover, the basis of albumen, fibrin, casein, etc., substances which differ both from their base, and from each other, in many of their physical properties. For example, albumen is soluble in water ; not so, however, fibrin and casein. Albumen coagulates under the influence of heat ; whereas fibrin spontaneously coagulates from the fluids in which it is held in solution, while casein is only precipitated from its solutions by dilute acids, and may be re-dissolved by excess of the same acids. These three bodies, thus differing from one another, and from their basis protein, are found to present the above chemical diversities; not because any such are exhibited by the protein, their great organic constituent, but because they vary in the quantity of their inorganic constituents, such as sulphur, phosphorus, sodium, chlorine, calcium, etc., mineral elements which are not contained in pure protein. It would thus seem that the varying quantities of these inorganic materials, in albumen, fibrin, and casein, are the influential causes of their respective physical qualities.

We have said that *chyme* is acid, and the *chyle* alkaline; and we have stated that from the salt of the blood, chlorine, or muriatic acid, is disengaged, as Dr. Prout supposes, by the immediate agency of galvanism; for he regards the principal digestive organs as a kind of galvanic apparatus, of which the mucous membrane of the stomach may be considered as the acid, or positive pole; while the liver, or hepatic system, may, on the same view, be considered as the alkaline, or negative pole; and he is of opinion that the greater part of the soda remaining in the blood, after the disengagement of the acid, is probably directed to the liver, and is elicited with bile in the duodenum, where it is again brought into union with the acid which had been previously separated from it. This view "illustrates the importance of common salt in the animal economy, and seems to explain in a satisfactory manner that instinctive craving after this substance which is shown by all animals." The great object of the digestive apparatus, as we have shown, is to prepare chyle from aliment, the general character and composition of which remains always the same. The stomach, therefore, is endowed with the power of securing this uniformity of composition, by an appropriate action upon the materials subjected to it. With respect to albuminous and oleaginous principles, the chief materials from which the chyle is formed, they require to undergo but little change in order to be fitted for reception into the system. But the saccharine class of aliments, as sugar, honey, starch, arrow-root, (composed of carbon and water,) which, excepting in purely carnivorous animals, enter largely into the food of mammalia and birds, are by no means adapted for such speedy assimilation. They have to undergo a chemical change, and become converted either into albuminous or oleaginous principles, and the stomach is a self-regulating apparatus, in which this conversion is affected. In the conversion of chyme into chyle, the pancreatic juice and the bile exert a decided influence. The chemical composition of the latter is very complex; it contains a peculiar resin, fat, or colouring principle, soda, salts of soda, mucus, and azotized animal substances, picromel, ozmazome, and cholic acid. Bile has the property of dissolving fat; and its bitter resin, which is highly stimulant and antiseptic, excites the secretion of the mucous membranes. True chyle cannot be formed without its agency; its constituents for the most part contain a large portion of carbon and hydrogen, which are obtained from the blood. Hence, while the bile is essential to the process of digestion, its elimination is one of the means for maintaining the purity of the blood. It is from venous blood, in minute vessels, termed venous capillary vessels, that bile is secreted, and it is the only known true secretion which is derived from this source, at least as far as the higher animals are concerned.

Such, then, are some of the discoveries relative to life and its processes which science has effected. Very limited, however, at the best, are the researches of human philosophy. Thoughtless men have sometimes objected to mysteries in the revealed word of God; but how much greater are those which exist in that department of the natural world to which our attention has just been directed! In it mysteries meet us at every turn. The phenomena of feeling, muscular motion, the antagonism between living bodies and the ordinary laws of chemistry, the conversion of food into blood, and the never-ceasing round of change which takes place in the body itself, are all so many mysteries. In the sight of God, then, let us be clothed with humility, and thankfully receive that revelation of his will which he has given us—a revelation in which he has made the way of redemption so clear that "the wayfaring men, though fools, shall not err therein." Science may be too deep for us; learning may be beyond our reach; brilliancy of intellect may be denied to us: but the truth that "God so loved the world, that he gave his only begotten Son, that whosoever believeth in him should not perish, but have everlasting life," is adapted to the capacity of all, from that of the lofty philosopher down to the humblest peasant. Here the sage finds himself on a level with the cottager.

CHAPTER II.

THE PURIFICATION OF THE BLOOD, ETC.

WHILE the blood suffers exhaustion in consequence of the demands of the system, it at the same time becomes vitiated, and requires purification; and this process is effected by the organs of excretion, namely, by the skin, the lungs, and the kidneys. These remove from the blood its vitiated particles, which, if retained, would soon induce disease. The suppression of any excretion is very dangerous; that of the lungs becomes speedily fatal, for in the latter case the carbon of the venous blood poisons the whole mass of the circulating fluid; the brain dies; the heart struggles feebly for a few minutes, and then ceases, and all the blood will be found to be black. If the excretion of the kidneys be suppressed, urea accumulates in the blood, fever supervenes, and coma and death follow. Let the secretion of bile in the liver be arrested, the same fatal result ensues; let its due excretion be prevented by any obstruction in the bile-duct, and disease, languor, incapability both of mental and bodily exertion, and indifference to everything, are the result. Again, let the insensible perspiration of the skin be checked, or suspended, and how soon the lungs and internal organs generally begin to suffer, and congestion or inflammation takes place. Let the excretion of the mucous membranes become greatly diminished, and manifold are the diseases which immediately supervene.

It is, however, to that purifying process in the lungs by which dark carbonized venous blood is converted into bright red arterial blood, that we would here more especially advert. This process is respiration. In animals with lungs, respiration consists of two acts, *inspiration* (performed in tortoises and frogs by a sort of deglutition) and *expiration*. Without entering minutely into the structure of the lungs, we may state that the bronchial tubes, or branches of the windpipe, pass gradually into a collection of minute vesicles, consisting of exquisitely fine membranes, over which the capillary branches of the pulmonary artery ramify; and that these membranes and capillary branches are permeable by air, which is brought into immediate contact with the blood. The pulmonary artery brings dark venous blood from the right side of the heart (the right ventricle) to the lungs; the pulmonary veins convey the purified blood from the lungs to the left side of the heart, (the left auricle, whence it passes into the left ventricle,) from which it is sent through all the arteries of the system.

At each act of inspiration, as the air rushes

into, and distends the lungs, they receive a tide of blood which fills all the capillary branches of the pulmonary artery spread over the walls of the thin air vesicles, and the air and the blood are thus brought into contact, and its purification is effected. At each expiration, the expansion of the lungs is greatly diminished, and a tide of blood, now arterial, and received by venous capillaries from the capillaries of the pulmonary artery, is propelled along the pulmonary veins to the left side of the heart. The quantity of air received into the lungs during ordinary inspiration is perhaps little more than a pint each time, but it may be increased naturally and without effort to two pints and a half. It must be remembered, however, that the lungs are never, in their natural state, exhausted of air, nor can any effort of expiration utterly exhaust them ; when, however, this is effected as far as possible, and followed by as forcible an inspiration as possible, a quantity of air, varying from five to seven pints, will be received. Upwards of nine pints have been so received, but this is beyond the average. About two ounces (by weight) of blood are received by the heart at each dilatation of the auricles respectively, and the same quantity is expelled from it at each contraction of the ventricles ; consequently, as the heart dilates and contracts four times to one respiration, or seventy-two times on the average in a minute, it sends every minute 144 ounces to the lungs, which in the same space receive about eighteen pints of air, in addition to ten or twelve pints constantly in the vesicles. The blood is estimated to perform three complete circuits through the body every eight minutes of time, and the average quantity of blood in the body, in health, is reckoned to be 384 ounces, or twenty-four pounds avoirdupois, being nearly twenty imperial pints.

We have now to ascertain the result of the action of the air upon the blood, and of the blood upon the air; and in following out this subject we shall see how the laws of chemistry are carried out, under the governance of a vital principle. Oxygen is the great agent in rendering the blood, whatever be the character of that blood, fitted for the purposes of the animal economy. No animal, whether it respires air or water, whether it is furnished with lungs or gills, or *vibratile cilia*, or whether the external surface alone acts as a respiratory organ, can live, unless a certain portion of oxygen be present in the fluid, whether air or water, which it respires. Every other gas in a pure state, such as azote or nitrogen, and hydrogen, and also carbonic acid gas, soon destroys the life of an animal forced by way of experiment to inhale it. Pure oxygen alone, although it will support life for a considerable period, is, when unmixed, too great a stimulant, and tends to exhaust the vital energies; oxygen must be diluted with azote; and of such is the atmosphere of our globe composed, with certain non-essentials, as carbonic acid, vapour, and exhalations from animal, vegetable, and mineral bodies. The water of the ocean, or of rivers, or of lakes, is by the atmospheric pressure replete with air, and is moreover composed of oxygen and hydrogen. Of one million of cubic inches of pure air, the weight of the oxygen is 71,809·3, of azote 238,307·7, total 310,117·0. Oxygen, then, is the food of life, and every animal destroys vast quantities of this food, insomuch that if the great vegetable kingdom did not produce oxygen, and fix carbonic acid, the atmosphere itself would in due time be incapable of sustaining life. As a proof of this, we may state that if any air-breathing animal, especially quadruped or bird, (for lower animals do not consume oxygen so rapidly,) be placed in a vessel of atmospheric air, so sealed up as to prevent all communication with the circumambient atmosphere, the animal, after a given lapse of time, perishes, and the oxygen of the air in which it was confined will be found exchanged for carbonic acid.

This leads us to the point in hand. The venous blood, having passed through the arterial system, and fulfilled the demands of that system, becomes replete with carbon, the presence of which renders it unfit for the purposes of life. Whence it acquired this carbon, or carbonic acid, is not very clear. "Some observations lately made," says Dr. Prout, "have induced us to believe that the conversion of albuminous matters into gelatine is one great source of the carbonic acid in venous blood. Gelatine contains three or four per cent. less of carbon than albumen contains. Now gelatine enters into the structure of every part of the animal frame, and especially of the skin; the skin, indeed, consists of little else besides gelatine. It is most probable, therefore, that a large portion of the carbonic acid of venous blood is formed in the skin and in the analogous textures. Indeed, we know that the skin of many animals gives off carbonic acid, and absorbs oxygen; or, in other words, performs all the offices of the lungs; a function of the skin perfectly intelligible, on the supposition that near the surface of the body the albuminous portions of the blood are always converted into gelatine." It is the presence of carbonic acid that gives darkness of colour to the venous blood, and on the removal of this it resumes its hue of scarlet. Carbonic acid is composed of pure carbon and oxygen, but the oxygen is not in a maximum proportion; and it has been demonstrated that between oxygen and carbonic acid a very powerful attraction exists. When, therefore, the atmospheric air is admitted into the lungs, the oxygen of that air unites with the carbonic acid of the blood, and, as experiments prove, is expired in the form of a gas, which will render lime-water turbid, and cause a precipitate of carbonate of lime. Thus freed from carbon, the blood becomes scarlet or vermilion, a colour due to the action of the salts it contains on the blood discs. It would appear, however, that the blood retains in itself a portion of pure oxygen, perhaps to become combined with carbon during its progress through the system.

According to the experiments of Dr. Prout, the generation and expiration of carbonic acid gas vary according to different conditions of the system, and also according to the hours of the day. For example, the lungs give out more carbonic acid during those of sunlight than those of darkness. The increase commences at daybreak, and at noon arrives at its maximum, decreasing as the shades of evening approach. With respect to the azote of the atmospheric air, it is returned unchanged, but generally diminished in volume, a portion of it having been absorbed into the system. With the carbonic acid and azote expired from the lungs, a large quantity of watery vapour is also thrown off, which appears, in a great measure, to be derived from the chyle which has recently been admitted into the venous system, and to constitute a means of its perfect purification. It is thus that the essential oil of various substances taken as aliment, and useless to the blood, although mixed with the chyle, is thrown out. For example, onions, garlic, rum, etc., impart their peculiar odour to the breath, and in indigestion, and during various diseases, the blood becoming loaded with deleterious particles, really poisonous, thus gets rid of them by a self-purifying process. The quantity of carbon thrown off from the lungs of a man during the twenty-four hours of the total day, has been estimated at about eleven ounces—a quantity, as Dr. Prout observes, more than equal to that contained in six pounds of beef. Nor can the generation of this quantity be easily accounted for, even on the supposition that it is owing to the conversion of albuminous matters into gelatine. A wide field of experiment is here open; the subject, however, is replete with difficulty.

With the function of respiration, according to most physiologists, the maintenance of *animal heat* is immediately connected. It is a certain fact, that on the union of carbon with oxygen heat is produced; and, as Dr. Crawford supposes, this production takes place from the union of carbon and oxygen in the lungs. Now, according to his theory, arterial blood has a greater capacity for caloric than venous blood; hence it follows that the heat generated during respiration in the lungs, by the combustion of carbon, is taken up by the arterial blood as rapidly as it is produced, and, consequently, it never becomes sensible, or raises the temperature of those organs above that of other internal portions of the frame. Were it not that the arterial blood was endowed with this capacity for caloric, the lungs would be destroyed or consumed by the combustion going on unceasingly within them. It must be acknowledged that arterial blood is sensibly warmer than venous blood, and perhaps the lungs are a shade warmer than other portions of the system. On considering this theory, which has been said to "afford one of the most beautiful specimens of the application of physical and chemical reasoning to the animal economy that has ever been presented to the world," we cannot say that we feel satisfied with it. In the voluminous lungs of tortoises and other cold-blooded animals, the same combustion goes on, but their temperature is little more than that of the surrounding medium in which they live. "It is exceedingly probable," says Dr. Prout, "that though the evolution of carbonic acid gas may be one of the means possessed by the animal economy for generating heat, there are yet other means, the nature of which at present is quite unknown."

We have already stated that more oxygen is absorbed into the system during respiration than is to be accounted for by the amount of carbonic acid expired, and we have intimated that this unites with carbon received into the blood during its circulation. It is in the capillary arteries that this union or combustion takes place, and that carbonic acid is formed. Now there is not a tissue, nor a portion of the body in which capillary arteries are not present in inconceivable multitudes; these arterial capillary vessels merge into venous capillaries, and transfer into them the carbonic acid which darkens the venous blood. From these venous capillaries it travels to the right side of the heart, and thence to the lungs, where a new combination with oxygen takes place, and the blood, losing its carbonic acid, returns renovated to the left side of the heart. Hence the capillaries of every part of the system are generators of caloric. In cold-blooded reptiles and fishes, the structure of the heart, and the condition of the arterial blood, are different from those of quadrupeds and birds. In reptiles, the heart consists of but one ventricle and two auricles ; and of the latter, the right auricle receives the vitiated blood returned from the system to the heart, the left auricle the arterialized blood from the lungs. Both auricles convey their contents into the cavity of the single ventricle. This ventricle (the interior of which, from the interlacement of muscular fibres, termed fleshy columns, or *carneæ columnæ*, assumes an almost spongy appearance) receives, therefore, both vitiated and arterialized blood, and these become more or less mixed together. Part of this mixed fluid is sent through the aorta to supply the system, and part through the pulmonary arteries to the lungs, to undergo a further degree of oxygenation—this ventricle having both the systemic and pulmonic arteries originating from it. Such is the routine of the circulation in the more perfect of the reptile class, namely, tortoises, lizards, and snakes. In the amphibia, as frogs, newts, the siren, etc., the circulation is that of a fish during the primary stage of their existence ; and in some, as the siren and proteus, continues to be such through life, although lungs are also given, but are in little actual use. In fishes, the ventricle receives the blood from the single auricle, and by its contraction transmits it into an enlarged arterial vessel, termed *bulbus arteriosus*, or arterial bulb, which soon divides into separate branches, one being destined for each leaf of the gills. Here the arterial vessels subdivide into fine capillaries, and these pass into branchial veins, which at last merge into two vessels, and these unite to form the aorta. Into this aorta, then, the blood, purified in the gills or branchiæ, is conveyed without first being sent back to the heart, and from this aorta it is distributed through the system. We must not here omit to notice the influence of the nervous system in maintaining the natural heat of the body. It is ascertained by experiments that the quantity of carbonic acid generated in the system and expelled from the lungs, is inadequate to the evolution of caloric, in proportion to the caloric abstracted from the body by the surrounding medium. It is proved that the nerves are the agents by which the due supply of caloric is effected; but the mode in which this supply is afforded is as yet not satisfactorily demonstrated. Some physiologists are disposed to think that the nerves have in themselves a direct specific power of generating heat; while others consider that the nerves operate indirectly, by the share they take in various processes of the organic economy. Be this as it may, the fact is established.

It will here be proper to make a few observations on two points, which demonstrate the energy of the vital principle, in reference to *temperature;* that of truly warm-blooded animals being especially under consideration. *First,* the body is capable of sustaining a temperature greatly *higher* than that of itself, and which would induce rapid decomposition in a body deprived of life. *Secondly*, when the body is exposed to a temperature considerably *below* that of its own natural or ordinary standard, the vital principle counteracts the depressing influence of that low temperature by the generation of caloric, according to the necessity of the case. But neither in the first nor in the second instance, as above adduced, can the vital energy struggle beyond a fixed point. The state of health, or a constitutional temperament, age, and other accidental conditions, must also be always taken into account.

Reverting to our *first* observation, it would appear that heated air, heated aqueous vapour, and heated water, raised each alike up to the same temperature, are not equally sustained by the body. The highest temperature which the human frame can struggle with is that of the air; and a temperature of 260°, or even more, may be borne without much difficulty, at least for a limited space of time. The temperature of aqueous vapour cannot be endured for more than half an hour if raised above 130°; although, as we are informed by Dr. Southwood Smith, the peasants of Finland are able to sustain it for the same length of time at the temperature of 167°. Here, most probably, custom has taught the system a lesson of accommodation. A water bath raised in temperature to 113° cannot be endured for above ten minutes, and then only by the robust, or by those inured to it. The reason why heated air is borne more easily than heated vapour or water, appears to be chiefly this—that from its expansion when thus heated, fewer ultimate particles come in contact with the body than would be the case were it at a lower degree of temperature. It must not be supposed, however, that the temperature of the body subjected to a heated atmosphere is not increased, for the contrary is the truth, but it is not increased beyond a certain point. Dr. Fordyce, in his experiments, ascertained that the heat of the human body, under the influence of a high temperature, speedily reached 100°, but that exposure to 211° did not raise it higher. Dr. Blagden, who was conjoined with Dr. Fordyce in a series of experiments upon animal heat, observes: "Being now in a situation in which our bodies bore a very different relation to the surrounding atmosphere from that to which we had been accustomed, every moment presented a new phenomenon—wherever we breathed on a thermometer the quicksilver sank several degrees. Every expiration, particularly if made with any degree of violence, gave a very pleasant impression of coolness to our nostrils, scorched just before by the hot air rushing against them when we inspired. In the same manner, our now cold breath cooled our fingers whenever it reached them; upon touching my side it felt cold like a corpse."* We know well how cold

* Phil. Trans. 1775, p. 118.

is produced by evaporation. The effects of the wine-cooler may be taken as a familiar illustration. The reason of this production of cold is as follows:—During the conversion of liquid (water, ether, alcohol, etc.) into vapour, caloric is absorbed, and by this absorption cold is generated, and the more so the more rapidly the conversion is carried on, insomuch that fluids may be frozen even during the heat of summer. Hence it is that ether feels colder than water from the rapidity with which it evaporates. Now, in the experiment of Dr. Fordyce above noticed, a rapid evaporation or conversion of moisture into most attenuated vapour, was taking place simultaneously from the lungs and the whole surface of the skin. The heat to which the body was exposed led to the vast increase of the natural exhalation which always takes place; and, therefore, the skin felt cold, and the breath caused the quicksilver of the thermometer to sink. Dr. Fleming denies that the body is kept cool by evaporation, because, when Dr. Fordyce was conducting his experiments in heated air, water poured down in streams over his whole surface; and he infers that this water was merely the vapour of the room condensed by the coldness of his skin, for at the same time a Florence flask, filled with water of the same temperature with his body (then 100°,) was bedewed with condensed vapour in a similar manner. Now this does *not* prove that exhalation did *not* perform a great part in the generation of cold, and we may easily conceive how *that exhalation* loading the air of the closed room became condensed on the surface of the body, which its evolution had tended to cool, as well as on the Florence flask.

And here we have another reason why immersion in heated water or vapour is not so easily endured as heated air, for during immersion in heated water, especially, the conversion of perspiration from the skin into invisible vapour cannot take place, and the work must fall upon the lungs alone. But there is another cause whereby the body is enabled to resist the effects of heat, and here we must look to the condition of the blood itself. During subjection to a high temperature, the pulse, or arterial action, is augmented both in power and frequency, and consequently more evaporable fluid is sent to the excretory tubes of the skin and lungs. But this is not all. Under the influence of heat the blood becomes less venalized, or deteriorated, and, in proportion as it preserves its fluid arterial character, the union of carbon and oxygen (that is, a gradual combustion) is greatly diminished, whether in the lungs or in the fine capillary vessels of the body itself. With this diminution of internal fire (for such it really is) is conjoined a diminution of temperature. We may state, then, as an axiom, that at an elevated temperature there will always be a diminished production of heat within the body, seeing that the blood contains a diminished quantity of combustible material, carbon, (or charcoal,) for combination with oxygen.

So far we have endeavoured to show why the body is enabled to struggle against a great increase of heat, within certain degrees of temperature; and we believe that the causes we have explained are all effectual, up to a given point. But we have yet to learn what part the nervous system plays in this counteraction; what galvanic or electro-galvanic changes are going on; and what is the agency of the brain, the spinal chord, and the intra-abdominal ganglia. As with the generation of animal heat, so with the generation of cold to counteract heat, there are powers within the animal frame which have hitherto eluded research, for the vital principle is in itself a mystery. Hence, much as we may learn, there is that remaining which can never be solved by human intellect.

Let us now attend to our *second* topic, namely, the power which the animal frame possesses of counteracting the influence of cold. Cold produces lethargy and death, yet up to a certain degree and for a certain duration of time it will be sustained by the system, according to the ratio of the health and of the vital energy of that system. There are repellent forces within the body, but these fail under the influence of time and degree—not, however, without a struggle. This struggle is maintained by the mysterious power within the organic frame, or, in other words, by vital energy. When the human body (or that of any warm-blooded animal) is subjected to cold, its temperature immediately sinks, but after the shock it rises by the agency of an innate force. Dr. Currie ascertained, (Phil. Trans., 1775, p. 121,) that the heat of the human body in one instance sank rapidly from 98° to 87° when placed in water at 44°, but at the end of twelve minutes it rose to 93°. In another experiment in water of the same temperature, the heat of the body fell from 98°, in the course of two minutes to 88°, but at the end of thirteen minutes had risen to 96°. "Dr. Hunter found that a dormouse, whose heat in an atmosphere of 64° was 81½°, when put into air at 20° had its temperature raised in the course of half-an-hour to 93°; an hour afterwards, the air being 30°, it was still 93°; at another hour afterwards, the air being 19°, the heat of the pelvis was as low as 83°, but the animal was now less lively. In this experiment the dormouse had maintained its temperature about 70° higher than the surrounding medium, and for the space of two hours and a half." In cold-blooded animals, heat is generated under similar circumstances. Hunter found that the heat of a viper placed in a vessel at the temperature of 10°, sank in the course of ten minutes to 37°; but in the course of ten minutes more, the temperature of the vessel being 13°, its temperature was 35°; in the next ten minutes, the vessel being at 20°, that of the reptile was 31°. In the case of frogs, he was able to lower the temperature to 31°, but beyond this point it was not possible to decrease the heat without involving the destruction of the animal. Even in the eggs cf birds, the vital struggle against destructive cold is vigorously maintained. Hunter found that a fresh-laid egg, (that of the common fowl,) put into cold water at zero, required seven minutes and a half more for freezing (and consequent destruction of vitality) than did another egg which had previously lost its vitality by freezing, but had regained the ordinary temperature by thawing. We might enter into a long detail of similar experiments were it necessary for our purpose, and, moreover, describe a series of trials, cruel as we think, upon the effects of the abstraction of caloric from the ears, limbs, etc., of animals, in order to test the extent of the restorative energy of the vital principle, but we have said enough to satisfy an intelligent reader.

Seeing, then, as we have endeavoured to demonstrate, that a body placed in a medium of lower temperature than itself attains nearly to its natural degree of heat after the first depressing shock, it will follow that the medium itself (whether air or water) must become elevated in its turn. Thus, a cold room is soon warmed, to use a popular phrase, by the assemblage of several persons, and so far from becoming a medium taxing the system for a generation of caloric, may, and often does, become one necessitating the generation of cold. And all this change, this adaptation to circumstances, this strain upon its vital energies, the animal system is ever bearing from month to month, and from year to year. Thus, too, in a cold bath, after the immersion of the body for a few minutes, the temperature of the water is decidedly elevated, and might be raised, by way of experiment, to that of the body itself. But how is this evolution of heat generated, as an antagonist against the effects of cold? It may be said, that the blood becoming more venous, that is, more carbonized, imparts its carbon, both in the lungs and in the capillary arteries of the system, to the atmospheric oxygen, and that the combustion thence ensuing produces a degree of heat sufficient for the preservation of the body. This theory, however, will not suffice. The ears of rabbits have been frozen, and have regained their natural condition—fishes have been frozen into ice-bound mummies, and their restoration has been effected—man has been frozen, his limbs being as insensible as the snow or the ice of the mountain, and yet, the heart being not quite dead, reanimation has been accomplished.

That the nerves have greatly to do with the struggle of the system against cold, is proved by the fact that the paralytic limbs of a sufferer (through which the circulation of blood is maintained) are colder than the rest of the body, and require the adjunct of artificial heat. Moreover, in certain experiments conducted by

Mr. (now sir B.) Brodie,* wherein respiration was kept up by artificial means after the animal—a rabbit—had been killed, the heat of the body was found to diminish as rapidly as in a dead animal of the same kind in which no attempts were made to keep up the respiration. Yet in the animal in which artificial respiration was carried on, the heart continued to beat for nearly two hours, the blood circu lated and was changed from arterial into venous blood in the capillary vessels, it was oxygenated in the lungs, and carbon was given off equal in quantity to that which is evolved in a natural state, and the oxygenated or arterial blood had the usual florid colour. It would appear that during the process of digestion, that is, the process of converting food into chyme, chyle, and blood, the system will vigorously resist the influence of cold. But what power of producing caloric can result from this ill-understood process? Ought we not rather to look for the cause in the action of those great nerves, especially the *ganglia* or *functional brains*, (the semilunar and cœliac plexus,) under whose agency the processes which conduce to the maintenance of the living frame in its organic integrity are continually carried on? In bodies cooled almost to death by exposure to cold, a bladder of hot water over the cardiac region, that is, over what is commonly called the "pit of the stomach," is very effective. Now in this region the *ganglia* or functional

* Phil. Trans. 1812, p. 378.

brains, to which we have alluded, are situated. It is because of these ganglia that a blow on the pit of the stomach fells a man to the ground, or even proves instantaneously fatal.

While upon this subject we may observe, that man and the higher animals have two species of life, one cerebral, or belonging to the brain, another dependent upon the functions of the ganglionic system. The first dies before the latter. The head may be struck off and yet the heart will beat, and the peristaltic action of the viscera continue, and, indeed, may by galvanic means, and by artificial respiration, be maintained for a considerable time. Were this not the case, a drowned person, dead as far as the brain is concerned, could not be resuscitated. In cold-blooded animals, as snakes, tortoises, etc., this ganglionic life appears to predominate over the cerebral. The heads of snakes and tortoises may be cut off, and the animals will crawl about and live for days—nay, even the vitality of the amputated head does not soon depart. We have seen a viper deprived of its head swim when thrown into water, and we well know that sea-turtles live for days after decapitation. But in these lower vertebrate animals, vitality and consciousness do not very quickly depart from the head struck off from the body, and we have even reason to believe that the head of man, or any highly organized animal, struck off by the axe or the knife of the guillotine, does not instantaneously lose consciousness, for the circulation of the blood in the brain for a moment or two must go on, and then fail, because no supply is afforded. Tortoises have been deprived of their brain, the skin has healed over the vacuum of the skull, and although the senses, such as sight, etc., derived immediately from the brain, were obliterated, the animals have lived and crawled about, dying at length, as it would appear, from inanition or cold.

Besides cerebral and ganglionic life, there is another kind of life, which we may call the *life of irritability*. We see this life where no nerves exist, and we see it also in muscular fibre deprived of all nervous influence. In the latter case, galvanism excites its display in animals recently killed. It is, however, most conspicuously exemplified in animals which, deprived of the head and viscera, move about, or in the creeping motion of the muscles of an animal slaughtered and cut up, quivering with unextinguished vitality. Some warm-blooded animals are more remarkable than others for this display of vital irritability. It is, however, in animals destitute of apparent nerves and muscles that its display most surprises us. We may adduce as examples the jelly fishes, (*Medusa*, *Linn.*) so common along our shores. These animals move and are sensitive even to the influence of light. The jelly-fish, by a sort of flapping motion, wends its way along the surface of the sea, and by means of pendant tentacles secures its prey; some species are formidable even to fishes, which are grasped and engulped. It is thus that the rhizostoma travels and supports its existence. The *Rhizostoma Cuvieri*, common in our seas, attains to a large size and weighs several pounds, and is often thrown upon the beach ; but when the water contained in its filmy cellular tissue is exhausted by evaporation in the heat of the sun, or becomes drained away, the mass of four or five pounds is reduced, as we have previously observed, to a sort of cobweb. Here the fluid contained in the filmy cellular tissue is in some mysterious way connected with vitality and the performance of organic functions ; and though no nerves or muscles are apparent in this strange structure, yet voluntary locomotion, contractility, and the power of preying upon fishes, are possessed. It does not, however, appear that the rhizostoma is endowed with *feeling* in the genuine sense of the word.

In the sea-anemones (*Actinia*) it has been asserted that a low nervous system has been detected ; certainly muscular fibres are present ; yet granting the existence of the former, light could not be felt by means of it ; nevertheless, it is light that invites these beautiful animals to expand like flowers, and adorn with their loveliness the rocks to which they are attached. A passing cloud, obscuring the rays of the sun, causes them to contract their many tinted tentacles, and shroud themselves in their outer tegument. We are here involuntarily reminded of plants. How many flowers, folded up during the night, and protected by the calyx, open on the dawn of day, or as the sun gathers strength! How many turn to the glorious orb, in its course from rising to setting! Yet they feel not,—theirs is a life of irritability, subject to the influence of certain stimulants. Venus's fly-trap, the sensitive plant, and others, afford examples of this species of vitality, which has obtained for them a kind of popular celebrity. The contractibility of celery, when torn or divided at the dinner table, is familiarly known to all, or rather to most who observe things around them. The life of irritability, then, is common to animal and vegetable tissue—manifesting itself by contractile or expansive movements; and we think that the predominance of this life is most marked, most striking, that is, it most arrests our attention, where nerves are indistinct, or altogether wanting. Thus several of the lower animals move, avoid injuries, seize their prey, expand or contract, and yet have neither brain nor apparent nerves; nay, the rays of the sun excite them, and some, if cut into pieces, become so many distinct entities. No one, we believe, will assert that even nervous molecules, setting aside nervous threads, are part and parcel of vegetable organization, and yet in vegetables the life of irritability is most conspicuous. Who that keeps plants in a window can fail to remark how they throw themselves towards the light? In common language, they are said to be "drawn" by the light; but in reality, they seek to gain the influence of the light. This impulsive movement is not confined to leaves or flowers; for it is a fact that roots seeking nutriment, by an innate vital force, which we cannot comprehend, and yet which must depend on the life of irritability, extend themselves in order to plunge into a propitious soil. Dr. Fleming says, "When the roots of a plant, spreading in search of nourishment, meet with interruption in their course, they do not cease to grow, but either attempt to penetrate the opposing body, or to avoid it by changing their course. Thus I have repeatedly seen the root of the *Triticum repens*, or couch grass, which had pierced through a potatoe that had obstructed its course; and everyone knows that the roots of a tree will pass under a stone wall, or ditch, and rise again on the opposite side, and proceed on their original direction." If we understand Dr. Fleming aright, this phenomenon is supposed by him to be under the governance of a species of instinct. We doubt this, however, unless, indeed, his definition of instinct be different from that of common acceptation. It appears to us that the improper, or uncongenial, and therefore irritating materials obstructing the roots, merely stimulate them to increased energy, for a root cannot hunt, as a hound or a mole, after its destined source of nutriment. It is stimulated, but has no aim, nor can it have aim, and the perforation of a congenial soil must be merely accidental. Does instinct, we may ask, direct the sunflower to turn its yellow disc to the orb of day, or teach the pretty pimpernel when to open and when to close, as the weather may be congenial or the contrary? Instinct, in our sense of the word, applies not to plants. It is not instinct that induces the window-plant to turn its leaves to the light, but its life of irritability is under the influence of the solar rays.

The example of vegetable irritability which is most accessible in this country, is exhibited by the barberry bush, so common in gardens and shrubberies. The phenomenon is thus described by sir James Edward Smith, who was the first to observe it. In the flower of this plant there are six stamens; these spread moderately, and "are sheltered under the concave tips of the petals till some extraneous body, as the feet or trunk of an insect in search of honey, touches the inner part of the filament near the bottom. The irritability of that part is such that the filament immediately contracts there, and consequently strikes its anther full of pollen against the stigma. Any other part ot the filament may be touched without this effect, provided no concussion be given to the whole. After a while, the filament retires gradually, and may again be stimulated; and when each petal, with its annexed filament, is fallen to the ground, the latter on being touched shows as much sensibility."

Another singular phenomenon of organic life must now be noticed—the *suspension of vitality*. Life, it has been well ascertained, may remain dormant or quiescent in organic bodies for a long time without producing any apparent effects beyond the preservation of the body itself — that is, without causing growth or development. This phenomenon is manifested in the eggs of birds, reptiles, and insects, etc., and in the seeds of plants. In order to render the vital principle energetic, certain stimulants and certain conditions are requisite, as a due degree of warmth, of moisture, etc., to which the body in question requires to be subjected. From the moment of that subjection a series of progressive phenomena commences, until full development is accomplished. Who, looking at the egg of a bird, and untaught by experience, would dare to predict that from it the peacock, long there immured, should become developed in all its glory? Who, looking at the egg of an insect, would, unless taught by experience, suspect that a gorgeous butterfly, or magnificent moth, lay there, sleeping in a little tomb? But because experience has taught us these things, the mystery is forgotten; yet the mystery still remains. Again, who, looking at a grain of wheat, would suppose that from its own decay it could afford nutriment to a vitalized germ, stimulated into activity by the properties of the soil, by moisture, and by the heat and light of the sun—blade, stalk, and ear, coming forth in their due season? Types these of man's great resurrection!

It may be received as a general rule, that unless a due stimulus be applied within a given time to these germs of organic being, sleeping but not dead, the vital principle languishes, and at length becomes extinct. But the time required for this extinction of vitality differs in different grades of life, and is preserved the longest, as far as we can positively determine, in the seeds of plants. It has been often observed, especially in America, that when a forest is cut down and the boles of the trees burned to the ground, a new forest (unless the ground be immediately cultivated) springs up, consisting of trees and brushwood, differing totally in species or genera from those which had previously grown there. It is a noted fact also, and one which we have ourselves witnessed hundreds of times, and never without interest, that when a bed of superficial clay is thrown up by the spade into mounds, hillocks, or ridges, it becomes rapidly covered by the common coltsfoot, and decorated with yellow blossoms. In both these cases, we believe that when the superficial deposit took place, the seeds, whence those trees, shrubs, or plants, are called forth, were at that time buried, and that they retained their vitality, waiting perhaps for centuries the day of vivification. That we have proof of the possibility of seeds retaining for two thousand years (under peculiar circumstances, not to be comprehended altogether) their vitality, and yet lying dormant during the revolutions of kingdoms, the extinction of languages, and the changes of the great seats of civilization, learning, and commerce—has been more than once or twice proved by the germination of grains of wheat taken from the mud-lining of the mummy cases of Thebes. The peculiarity of the wheat bears in its character evidence of its antiquity. It is the wheat of mighty nations, high in civilization, whose power and splendour, as predicted by the prophets of old, passed away, or barely lingered, before what we call antiquity commenced.

What is this long continuance of dormant vitality, but a modification of that law which governs the torpidity of so many animals in what is commonly termed a state of hybernation? The seeds of plants torpid for centuries under ground, have still a latent life—they have never parted with the vital principle. The same observation applies to certain animalcules, (*Rotiferæ*,) which after years of latent life have been restored to activity. Who then can tell through how long a period the germs, infinitely minute, of these animalcules, endowed with latent life, may be carried along by the wind, swept over continents and oceans, buried in the sand of the desert, or entombed in the earth, waiting, so to speak, for a proper *nidus* or medium of development?*

* Marvellous indeed are these phenomena, but they are not without a parallel even in the moral world. We have alluded to the vegetation of plants, long buried under the soil, and quickened into life by being unexpectedly turned up to the earth's surface. So often it has happened on the stage of history. Individuals who had remained in quiet obscurity, have had talents and properties of good and evil, previously unsuspected by themselves, and by all around them, rapidly developed by critical emergencies. Nor even in the spiritual world are we without kindred analogies. As seeds vegetate after long appearing destitute of vitality, so often have instructions and counsels, communicated by pious parents or faithful pastors, become—through the Holy Spirit's quicken-

Another interesting phenomenon now requires a brief notice at our hands; the process by which such injuries as are not necessarily destructive to life are more or less completely *repaired*. Thus, for example, a cut wound heals in due time, although a cicatrix, or seam-like mark, remains to testify of the previous mischief. When a branch is rudely torn from a tree, the bark around the wound takes upon itself a curative process, and accumulates; sometimes the edges of the wounded part meet and unite, but generally the bark forms a thickened ring, leaving an exposed part of pure woody fibre to perish. The reason of this arrangement is obvious. The woody material is to the tree a non-essential, excepting in so far as it is requisite for stability. All the great vital functions of the tree are carried on in the bark, hence the vital energy of the bark is primarily directed to the cure of its own laceration. It is a singular fact, that as we descend in the scale of organic creation, and pass from animals with a highly developed brain and nervous system to lower groups, we find the energies of vitality in the curative or restorative processes of wounds and mutilations more and more

ing grace—vivified after the lapse of intervals of time, more than sufficient, it might be supposed, to have destroyed all traces of them. What a motive should this be to all who are engaged in offices of Christian philanthropy never to faint through apparent want of success! The seed sown and apparently lost may turn out productive after the lapse of many days. Were facts of this order carefully collected, we might find numerous parallels to the well-known anecdote of the man, who, when nearly a hundred years old, was brought to a saving knowledge of the truth by the recollection of a sermon which he had heard when a mere youth.

influential. In the higher orders, indeed, fractured bones are reconsolidated, divided tendons reunited, severed muscles re-agglutinated; but here the restorative power ends; no amputated limb can be recovered. Among several groups of lizards,—and we may instance the geckos,*—the tail, from the character of the articulations, and a strange peculiarity of the fibres of the muscles of the part, is exceedingly brittle, and snaps off with a slight touch. It is, however, soon replaced by a new tail, but a swelling at the base of the reproduced member, like the swelling of the bark around the laceration of a tree, marks the line at which the new growth commenced. The tail of the common viviparous lizard (*Zootoca vivipara*) of our sunny banks and thickets, is also extremely brittle, and when broken off, is in due time restored. Lobsters and crabs present us with examples of animals capable of reproducing their limbs after forcible removal; nay, the lobster, when excited by fear or some other impulse, will, by means of a sudden jerk, throw off one of its great claws, and thus for a while lose one of its natural weapons. In a short time a new claw begins to *bud*, and becomes at length fairly developed, but does not acquire a new covering until the periodical renewal of the whole shell takes place, when the limb, together with the rest of the body, acquires a renovated suit of armour. It is probable that in spiders this

* See Popular History of Reptiles, Religious Tract Society p. 126.

reproduction of limbs also takes place; but in true insects, generally speaking, their life is too short to admit of the accomplishment of such a process, although that of healing, if not of restoration, is effected.

But the mere restoration of limbs is a trifle; it would seem as if it were a benefit to some animals to sever them asunder, inasmuch as two or more distinct creatures are the result of the operation. For example, in the sea-anemones (*Actinia*) the irritability, or, as some please to term it, sensibility, is so great that they will contract their tentacles, and fold up their flower-like forms, even when a dark cloud passes over them; yet such is the vital energy of these beings, that when cut asunder transversely, each part will become a perfect and distinct animal. In this case the basal part is about two months in regaining tentacles. They may be also longitudinally divided with the like results. Certain small leech-like animals, termed *planaria,* which inhabit both fresh and salt water, are not only multiplied by division, but even spontaneously separate themselves into distinct portions, each portion becoming an independent being. To the sea-stars also (*asterias*) this power of self-multiplication is abundantly given. If torn into pieces, each fraction becomes a distinct animal, ready to carry on its destructiveness among the beds of oysters. It would be easy to adduce multifarious examples of this power of reparation of injuries, and of distinct individuality resulting from the positive division of the frame of many of the lower orders of the animal kingdom. But further examples are not needed; and yet, as we are dwelling upon the energy of the vital principle, we can hardly pass over the astonishing fact, that among certain of the lower orders of creation, namely, the polygastric animalcules, reproduction is by buds, by spontaneous division, and by eggs, while some species exhibit each of these modes of continuing the race.

CHAPTER III.

ORGANIC AND INORGANIC MATTER COMPARED.

We shall now proceed to consider organic bodies, as contradistinguished from inorganic bodies.

Every animal and every vegetable, we may remark in the outset, has its restricted form. This, it is true, can also be said of crystals, and the beautiful crystallization of moisture on the panes of glass in our window during a sharp frost, may be cited as a familiar instance in point. But crystals have not their forms according to the laws of vitality. A rich arborescence may be mimicked by them on the frosted window-pane, but the illusion vanishes with a slight increase of atmospheric temperature. The atoms were not bound into one living thing by a vital principle. The fixed rigidity of crystallization is never, we may remark, seen in organic bodies, even when these bodies are most rigid in their lines and contour. Crystals, too, have a homogeneous composition. Organic beings, on the other hand, are composed of heterogeneous substances and parts, of solids and fluids, between which changes are perpetually taking place. Change, in short, is one of the grand characters of vital organization; while permanence is that of a crystal, till its form is destroyed by an extrinsic, not intrinsic, agent.

Organic beings have all, more or less palpably developed, a central form, to which are attached members, diversely modified, but essential to their well-being. The exceptions to this rule, as in the case of the sponge—if we can include it within the animal kingdom—and of certain gelatinous corpuscles, regarded as coming within the domain of the vegetable world, need not here be taken into the account. On the whole it may be said, that animals as well as plants have "a more or less rounded or cylindrical form, branched or membered, bounded by curved lines, and by convex or concave surfaces, very distinct from the crystalline, the only regular form of inorganic matter." In all animals, the form is definite even when precise symmetry is wanting; in most, however, form is symmetrical as well as definite. Our distinction may be thus illustrated—bivalve mollusks, as the oyster, are definite in form, but not symmetrical; the nervous ganglia, the limbs, or appendages, the aerating organs, etc., do not balance each other in and around a central mass. But in the classes of quadrupeds, birds, reptiles, and insects, as well as crustacea, (crabs, lobsters, etc.,) the nervous system, whether a true brain be present or wanting, is symmetrical, the limbs are symmetrical; for example, mammalia have four limbs, two on each side of a central body, or, as in the whales, only one on each side; birds have two wings and two legs, symmetrically disposed; many insects six legs, and two or four wings, or two wings with wing-covers, also symmetrical. In crustacea, the forms of which are often most fantastic and extraordinary, the same rule holds good. To make our meaning still clearer, let our reader place before him a butterfly and an oyster, and he will then have an example of definiteness of form in both, but of symmetry besides definiteness in the former. In the vegetable kingdom, there is a greater latitude both as to definite and symmetrical form than in the animal kingdom. These observations lead us next to the consideration of what is termed *specific form.*

Every species reproduces its like; but every species is subject to variety, even including man, the highest of the animal creation, who varies in stature, complexion, and language, throughout the great divisions of the globe. It is, however, to the vegetable kingdom that we would here rather apply what we particularly mean by specific form, because no two trees of the same kind are exactly alike, no two bushes are alike, no two plants are the similitudes of each other. Yet in one sense they are respectively alike; for who would mistake the oak for the ash, or the palm-tree for a pine? After its "own kind" the tree springs forth from the seed, and the vital principle fails not in the fulfilment of its purpose. Hence we may predict with certainty, that from the acorn will the sapling of the oak spring up, or the sycamore from its two-winged seed. So, also, from the egg of the eagle will the eagle be produced, and from that of the dove, a dove. From the egg of the butterfly the caterpillar will emerge; this caterpillar will become a dormant pupa; and from the pupa case, soon to be riven asunder, shall emerge the sportive butterfly, painted according to its species, and which species alone it can reproduce. All plants have their own peculiar forms, and these forms are as characteristic as are those of animals. They are *specifically* characteristic. The oak, the birch, the willow, and the pine, present us with features, irrespective of leaves, which distinguish between them. Yet no two trees of either species shall be similar to each other in the number of their larger branches, the precise outline of the trunk, or the point at which ramification commences. With respect to the smaller branches and twigs, still further differences will prevail; and yet throughout the whole, from the root and stem to the minutest ramification, specific distinctiveness will manifest itself. Specific form, then, in organic beings, is the result of a vital principle, not of chemical laws; but at the same time within *that form* many chemical phenomena are in unceasing operation, subordinate to a higher power. Herein there is mystery; for how can we explain the working of organic and inorganic laws in mutual co-operation, and yet, without clashing one against the other, each operating in its own province to the accomplishment of a *living thing?*

The vital principle determines *form;* not only *form,* indeed, but *size* or *magnitude,* is bounded by the operation of this principle. No animal or vegetable exceeds beyond a very trifling degree the normal stature, bulk, or weight, of the generality of its species. The rose will not attain to the magnitude of the oak, the myrtle to that of the cedar of Lebanon; man cannot attain to the bulk of the elephant, nor the frog (though he were ambitious of the feat, as is wittily represented in a wise fable) to that of the ox. To the microscopic animalcule the gnat is a giant—the butterfly, a monster.

Generally speaking, growth (in each species after its kind) is progressive, from the dawn of life or immaturity up to a certain point, at which further growth ceases. This we call maturity. This growth is more slow or more rapid, according to the species; but why it should cease at a given time we cannot tell. It does so, and that is all we know. To the law of the increase of growth from the dawn of life to maturity, there are certain exceptions—perhaps more apparent than real, inasmuch as the fluid portions of the body are in excess, until an adult condition be attained. We here more particularly allude to certain races of insects, as butterflies, moths, etc., the caterpillars of which weigh more, and are in fact larger at a certain stage of existence than will be the pupa, or the perfect image; but then they are replete with fluid, from which the organism of the perfect insect in a more concentrated form is to be produced. The caterpillar feeds voraciously on the leaves of plants, and makes havoc in the garden; the butterfly sips the honey of the flower.

The duration of the natural life of animals and of vegetables is very different in different species; but here, also, some peculiarities are observable. For example, some insects live for one, two, or three years in a larva state, but soon die when the perfect form is assumed, the female having time only to deposit her eggs. In this point of view the ephemera is not, as generally supposed, the being of a single day.

As among animals, so among plants, the duration of life is very variable; some, from the germination of the seed in spring, endure only to the close of autumn. To others a life of two or three years is allotted; while others attain not to maturity under centuries, and at length decay—the wrecks of ages. Yet some trees may be said never to die. The banyan-tree, (*Ficus Indica*,) forming a limited forest by the "rooting into the earth" of its dropping suckers, multiplies its own existence, and as each newly-formed tree is in more or less immediate communication with the parent trunk, so this trunk, the patriarch of a vegetable nation around it, perpetually renews its vigour.

The celebrated banyan grove of India, which gave shelter to the troops of Alexander, still exists; it affords covert to wild beasts below, while tree-snakes and hordes of monkeys find security amidst its umbrageous ramifications. It may be called a forest composed of a single tree, under the canopy of which oriental and European troops have in modern days encamped. Many travellers have described it, but never without expressions of delight and wonder.*

Decline succeeds maturity, and death is the ultimatum of life. Inorganic matter, however, neither declines nor dies. A crystal may be dissolved, or become converted into a different kind of crystal by accidental or designed chemical agents; but whatever change it undergoes, no vital principle is lost, because no such principle was ever possessed. Hence a crystal, isolated from external agency, will preserve its integrity for ever. A crystal may be increased in mere size by layers upon layers of external addition; but an animal or vegetable adds to its own growth by a digestive process, wherein the nutrient particles are changed from their original condition, and become part and parcel of the organic body. No inorganic body is endowed even with irritability, yet many exhibit peculiar electric or galvanic properties. Here, however, animal substances present a counterbalance. Many vegetable and animal

* See Oriental Annual for 1834; and Asiatic Researches. iv. 310. Forbes in Oriental Memoirs.

bodies, whether living or in a state of decay, display the luminosity of phosphorescence. Among plants, the marigold is an example in point; while in the animal kingdom most, if not all, of the lower marine animals are luminous. We have seen the shore studded with innumerable stars, each luminously blue, giving out from itself a decided pale blue light. On examination, these stars were a species of beroë. We need not here advert to the glow-worm, nor to the effulgence of the ocean in warmer latitudes—an effulgence resulting from the congregated myriads of marine animalcules, *crustacea* or *acalephæ*, spread out over leagues of the briny deep. The flash of the diamond, and the glowing effulgence of animal phosphorescence, have nothing in common with each other.

While organic matter increases, attains to maturity, and declines, it produces inorganic matter; nay, even crystals, which, once formed, are governed by the laws of inorganic matter. Silex, or flint, secreted by the plant itself, forms the exterior incrustation or bark of grasses, reeds, and canes (*Endogens.*) Nay, even spicules and crystals of silex have been discovered within the hollow of cane joints. We have often heard persons of some information ask, how, and from what source, a plant, rooted in a soil destitute of silex, can acquire this deposit; but *silicium*, or *silicon*, is a chemical basis everywhere abundant, and when united with oxygen the product is silex, or *silica* (a non-classical word.) It is therefore by its own internal vital powers, involving chemical phenomena, that the endogenous plant coats itself with silex. May not this process be called *vitali-chemical?*

All organic beings, after a certain period, begin to lose a portion of their vital energy. Animal heat is then maintained with difficulty, and the nervous system loses its functional delicacy; the system flags, the arterial pulse is no longer bounding, and the black blood overladen with carbon creeps torpidly through the veins. What applies to the condition of animals applies also to vegetables. In these organic forms, whether destined for a brief or an extended life, the circulation of the sap (their blood) first begins to flow languidly, till at length the conclusion is death. This observation applies to the death of a tree in its integrity, or to the death of every leaf which has unfolded in spring, seeing that every leaf is a distinct being, dying not before a successor is prepared, while at the same time every leaf is in vital union with the living bark through which the circulation of the vital fluid is maintained. How closely the tree and the compound zoophite assimilate!

Another interesting peculiarity connected with the energy of the vital principle now demands our attention. A superficial survey of the animal kingdom leads many persons to believe that an individual of any given species is an isolated being, or rather, a being in itself independent, whether it be male or female.

But such a survey will not tend to our knowledge of nature. Superficiality and error are almost convertible terms.

We shall not here enter into the physiology of the combination of male and female in the same individual. Among the lower orders of the animal kingdom, and among plants in general, (with certain exceptions,) this arrangement, for obvious purposes, has been ordained by the Creator. Passing from this topic, we come to another of no little interest, namely, the union of what may be termed distinct centres of vitality, so as to form a composite whole. Here we would first allude to the zoophytes, (class *Phytozoa.*) On our own coast many beautiful specimens of these plant-like beings exist, as *Thuiaria campanularia,* and *Tertularia;* it is, however, in the warmer latitudes of the ocean that the phytozoa display their most astonishing variety of forms; some appear like fans of network; others, like pliant oziers, bend before the impulse of the current. Here rise up corals with chaliced flowers; there madrepores, encrusting rocks, form reefs dangerous to the navigator, or rise in islets covered with intertropical vegetation. The zoophytes are polypiferous beings, and though in general each polype forms only one of an assemblage, united together by a common tie—as the bark unites into oneness the whole of the foliage of a tree—yet each polype acts for itself independently of the rest, and expands or closes its floweret-form, according to the stimulus which immediately affects it. Here, perhaps, by way of rendering the matter clear to our readers, the author may be permitted to quote from his own writing, making such omissions and alterations as are necessary for brevity's sake. "In order to convey an idea of these polype-bearing animals, let us begin with the simplest of their forms. There is in fresh-water ponds or slow rivers, a minute gelatinous creature termed *hydra*, of which several species are known. These animals are of slender figure, with an internal cavity, and a mouth surrounded by movable arms or tentacles; they possess the most extraordinary powers of elongation and contraction, and are highly carnivorous; they attach themselves, by means of a caudal sucker, to the leaves of aquatic plants, and spread abroad their arms in quest of prey, which they drag to the mouth and engulp. No nerves or muscular fibres have been detected, but the gelatine composing the body is replete with minute granules. The hydra is tenacious of life, and when cut asunder each part becomes a perfect and independent being. Yet, though apparently insensible to pain, the hydra appears to appreciate the presence of light, and its tentacles doubtless feel the prey around which they cling. The hydra is free, and moves about over the surface of the leaves, or swims in the water with the tentacles downwards, the caudal sucker acting as a float. When alarmed, it contracts its tentacles, and shrinks into the form of a small globule, easily escaping observation. Here we have an example of a free, independent, gelatinous polype, of the simplest structure, the digestive apparatus being a simple excavation in its substance ; it is in fact merely a digestive sacculus, the orifice of which is fringed around with prehensile tentacles. Now suppose one of these animals should greatly expand, and deposit within itself a calcareous substance, acting as a sort of rude skeleton, and hence become, though not fixed, incapable of locomotion—such a being we have in the *Fungia*, one of the *Madrephylliœa*. In this creature the stony axis has the upper surface adorned with radiating plates, rendering it not unlike a mushroom. The species of fungia are found in the Indian seas, and lie loose and unattached upon the soft sand at the bottom of the water. The gelatinous investment is contractile, and reproduction takes place by buds or gemmules, which at length become detached, and are carried away by the waves."

Let us advance another step—let us picture to ourselves a gelatinous extension common to many polypes, united as it were into one compound unity, and secreting for the internal support of the whole, a branching calcareous tree, with pits or cells on the branches, in which the polypes individually reside, and from which they may protrude and expand their tentacles. Examples of this form we see in many madrepores, as *Oculina*, *Dentipora*, etc. In other instances, the gelatinous expansion may secrete large calcareous masses for its support, with polype cells variously arranged. In these instances the calcareous support is firmly consolidated to the rock on which it is based, and the polypes, often of most beautiful tints, resemble thickly-scattered flowers on a glistening bank ; as examples in point we mention *Astræa, Meandrina*, etc.

It not unfrequently occurs, that the axis, or skeleton, invested by gelatine is not firm and calcareous, but horny and elastic ; in this case it is generally in the form of a leaf, or fan, or it resembles the branch of a weeping willow, and, rooted on the rock, bends to the waves. In these instances there are not always polype cells in the substance of the horny stems or twigs, but the cells are placed in the gelatinous bark, or cortical substance, investing the stem and twigs, and within these cells the polypes or animal flowerets dwell. In some forms, as in *Iris hippuris*, the stem is composed alternately of joints of flexible horny matter, and of beads of white calcareous matter, the whole being invested by a gelatinous bark of considerable thickness, in which the polype cells appear like numerous pits or dimples.

In the cases cited above, the living gelatine secretes or deposits an internal solid substance, as the inner bark of a tree secretes the wood, which is the common mechanical support of the whole. But we must now reverse the picture, and imagine a number of polypes united by thread-like films into one being, these threads being cased with a tubular sheath of horn, which they themselves have secreted by way of support and defence. "Let us picture the whole as a tuft of delicate vegetation, a frondescent plant of a tubular horn, with orifices, or cups, or bells, bestudding every branch in which the polypes reside, and from which they can protrude their contractile tentacles. Such have we in the *Sertulariæ*, compound tubular *Phytozoa*, which we find rooted to stones and shells in abundance on our own coast." "There are, however, some *Phytozoa* which inhabit calcareous tubes, and in some instances these tubes are collected into masses of considerable extent, all ranged in order, like reeds bound together, or the pipes of an organ. They are open at one end, through which the polypes, independent of each other, except when they form the bands or stages which unite the pipes together, protrude their flowerets. Such is the beautiful *Tubipora musica* of a deep red."

Distinct alike both from the cortical and the tubular *Phytozoa* are the *Alcyonidæ*. In these we find neither a horny or calcareous axis, nor a horny or calcareous sheath, but are here presented with a firm cartilaginous mass, having calcareous *spiculæ*, at least in some instances, dispersed through its substance. This mass is studded with flower-like polypes, each in its own cell, the cells being excavated in the living gelatinous mass, to the nutriment of which, as the common bond of union between them, they all contribute. These may be called compound cartilaginous *Phytozoa.* We might here enter very largely into the manifold forms which the zoophytes present; our aim, however, is only to convey some idea of the general character of a group of beings within the pale of the animal kingdom, which exhibit a vital union of distinct beings, each of which, acting for itself, contributes to the common good; for truly in one sense we may consider each polype as a distinct being; in another sense, as forming a part of a compound unity, through which vitality is equally diffused. In these aggregated forms the polypes all labour to one end; they constitute a community, every individual of which contributes to the sustenation of the general body. But as there are no nerves, they cannot participate in each other's movements; nor does any stimulus which affects a single polype, influence all the rest at the same time; if one polype be destroyed, the rest are unaffected. In the consideration of these beings, all ideas of life, derived from a study of the higher orders of creation, must be banished. "Deadness to pain, yet sensibility to light; contractility, expansibility, and motion, with muscles; digestion and nutrition without apparent absorbents or blood vessels; reproduction by eggs, by simple division, or by bud-like sprouts; the vital unity of myriads, and yet their personal distinctness—these are characteristics, which surprise the more, the more we reflect upon them. When to these characteristics, also, we add those taken from their rude skeletons, their horny tubes, their cells, their plant-like forms, and mode of attachment, we must confess that the *Phytozoa* are among the most mysterious of living things which the fiat of creative Wisdom has called into existence." To these strange beings the term *Phytozoa* (from φυτον, phyton, *a plant*, and ζῶον, zōon, *an animal*) is very applicable. As, from a single seed, a tree or shrub, with all its branches, leaves, and flowers, is developed, so from a single ovum does a branching zoophyte, with its numerous cells and polypes, become developed. Among these cells are some which may be termed egg-capsules, in which the ova are developed, and after a certain time escape in the form of ciliated bodies, which being dispersed ultimately become fixed, and grow up into branched zoophites, or corallines. " In some species the germ cells, or egg capsules, are metamorphosed into ova at particular parts, and the concomitant growth of the soft tissue and outer crust furnish these ova with a capsule; which modification in the growth of the coralline, professor E. Forbes compares with that metamorphosis in flowering plants in which the floral bud is constituted through the contraction of the axis, and the whirling of the individuals borne on that axis, and by their transformation into the several parts of the flower." Trees and plants like corallines exhibit an aggregation of distinct individuals vitally united into one whole. Professor E. Forbes says, " We are not in the habit of regarding the leaf as the individual

—popularly we look upon the whole plant as an individual. Yet every botanist knows that it is a combination of individuals, and if so, each series of buds must certainly be strictly regarded as generations."* Professor Owen observes, that "Both plant and zoophyte proceed to develope by gemmation—the one a succession of leaves, the other of polypes, associated by the continuous growth of the connecting parts; and finally, the plant by metamorphosis of part of the stem and certain leaves, produces the flower or fructification, and the zoophyte, by a modification of its stem, and certain polypes, produces an egg-vesicle, or a modified polype, or a medusiform individual, which is set free; in both cases the end to be attained is the diffusion of the species by means of impregnated seeds or ova." The learned professor places before us the grand characteristics of plants, as organic beings, in a striking, and we may add, original manner; but our prescribed limits prevent us from following him.†

Plants exhibit several modes of continuing their respective species; some propagate only by seeds, which the wind scatters abroad, and which, dropped in favourable spots, germinate and grow. These seeds are vitalized in the pericarp, capsule, drupe, or seed vessel, whatever be its character, from which in due time and by various methods they are liberated.

* Magragraph on British Naked-Eyed Medusæ.

† See his Introductory Discourse to the Hunterian Lectures, 1849.

Some plants propagate by gemmation as well as by seeds. They throw out suckers, as we commonly term them, from the root, and thus are stems multiplied, each having its own root. Of this the rose is a familiar example. Some, after flowering, die down in winter, while, on the return of spring, gemmation takes place in the roots, and verdure is renewed. Of this the garden-mint, the common nettle, and others, are examples. Bulbous roots, as the hyacinth, renew their leaves, stalk, and flowers in spring; but such roots multiply by a sort of spontaneous division, young bulbs budding from the parent root, and increasing in size until a final separation takes place. The potatoe and the Jerusalem artichoke propagate by seeds, and also by tubers, which form on the extremities of the roots; the plant dies, but these tubers remain alive, and gemmate, more or less abundantly; each *eye*, as it is commonly called, being the speck of gemmation. Hence from one tuber numerous plants are produced; each plant in turn produces many tubers, and each tuber again many plants; and thus the species becomes multiplied. Some plants, with trailing branches or stems, strike roots into the earth, wherever the earth and stem, or branch, are in mutual contact. Hence, from one small plant a large bed will result, as in the case of the garden sage. Carnations are thus artificially multiplied. Garden pinks abundantly multiply themselves by self-struck layers. The Indian fig forms a forest by sending down suckers from its spreading branches, which on touching the ground become rooted. Many of the lower animals, as we have seen, such for instance as the planaria, the sea-anemone, the sea-star, or asterias, become multiplied by artificial division. In like manner numerous plants may be similarly multiplied. Take, as familiar examples, the willow, the myrtle, the scarlet geranium, (or *pelargonium*,) the cactus, the pink, and others, which will suggest themselves to the mind of our reader. The difference is this, that plants require to be rooted into the moist earth, whence they chiefly derive their nutriment; while the animals to which we have alluded have a self-healing power, as well as the power of developing new organs, provided only that they be not removed from their natural element. No animals are ever rooted; and yet the plant-like corallines which we see attached to stones or shells seem as if they were rooted, but in these instances the root is only a root in appearance. When the minute germ became attached to the stone or shell, it began immediately to send out fibrils of attachment, and attachment only; not for the purpose of taking up the nutriment, for this is to be effected by the polypes, as the stem with its external or internal support becomes developed. In those large species from the warmer seas, in which a thick, firm, granular gelatine invests, like a rind, the calcareous or horny support, this rind spreads out at the base, over the stone, or madrepore, on which the whole is fixed, insinuates itself into every crevice, conforms to its superficies, and cements the first layer of the internal skeleton to the substratum on which it is deposited, and from which it must be broken off. We are here reminded of the growth of the antlers of the stag or fallow deer; the burr of which is rooted on the skull, and the stem and branches of which are secreted by a velvet-covered skin, which in due time perishes. There is, however, this difference; the velvet of the antlers is a peculiar and temporary development of the skin of the animal, in order to effect a given object. In these fan-like, or frond-like zoophytes, the investing and secreting rind, or integument, results from the development of a minute germ, alien to the rock or stone on which it chances to fix. Yet the analogy between the growth of a fan-like zoophyte and the antlers of the deer is very striking, and a consideration of the process in the latter case will lead the reflective mind to form a clear idea of the process in the former; only that in the case of the zoophyte, secretion of horny or calcareous matter is not carried on by the agency of distinct and palpable blood-vessels.

To revert to plants. There are many species which never bear flowers—never exhibit a calyx, petals, stamens, or pistils, and yet multiply by prolific seeds. We may here instance the ferns. In these plants the seeds are arranged at the back of the fronds, either near the summit or at the base; and this fructification is either naked or covered with a membranous envelope. In *Lycopodium* the fructification is axillary. In the *Hepaticæ*, or liverworts, the herbage is frondose, and the fructification originates from what is at the same time both leaf and stem. The seeds are contained in capsules. In the *Algæ*, the herbage is frondose, and generally of a leathery or gelatinous texture. There are no flowers, but the seeds are imbedded either in the frond itself or in some peculiar receptacle. In many of the marine *algæ*, the seeds are lodged either in external capsules, tubercles, or vesicles, or in the joints of the frond. The submerged *algæ* are in general merely fixed by the roots, their nourishment being imbibed by the surface, and many of them float without being attached to anything. Herein we are reminded of those frondescent corallines, the roots of which are merely fibres of attachment. The *algæ* are either jointless, as *Fucus vesiculosus*, a common seaweed on our coasts, and the laver, (*Ulva*,) esteemed for the table; or jointed, as the *Confervæ; or disjointed*, as *Diatoma*, *Fragillaria*, etc. When the jointless *algæ* fructify, the fronds either develope little cellules, in which the reproductive grains are inclosed; or some part of their cells changes its appearance, acquires a deeper colour, and finally drops to pieces; or the whole mass of each individual, at a certain stage of existence, appears to separate spontaneously into particles, each destined to become developed into a distinct plant.

The jointed *algæ* or *confervæ* are chiefly inhabitants of fresh water. They present the appearance of thread-like tubes, having joints, differing in length and the manner in which their contents are arranged. An endless variety of these plants may be found in clear ditches and running streams. They multiply by means of little granules contained in their tubes, and they grow by the addition of one tube to the the end of another. Among these *confervæ* the most remarkable are the *Zognema* and *Oscillatoria*, both of which evince certain degrees of approach to the animal kingdom. "The species of the latter genus form dark green and purple slimy patches, in damp places or in water, and are exceedingly remarkable for the power they possess of moving spontaneously; when in an active state their tubes are seen to unite and twist about, just as if they were vegetable worms; but they grow like plants, and their manner of increase is also vegetable. Yet they possess several of the chemical properties of animal matter, and when burnt yield a carbon of the most fœtid odour, resembling that of decaying animal substances." Have we not here a link of union between the zoophyte and the vegetable?

Disjointed algæ are extremely curious; they are characterized by their original or final spontaneous separation into distinct fragments, which have a common origin, but an individual life. They multiply by spontaneous division. "It is upon the stems of other plants immersed in water or floating in pools or ditches that these curious productions are met with; in their habits they are so paradoxical, that naturalists are far from agreed as to whether they are not really minute animals, but their mode of growth seems to compel one to answer such a question in the negative. At the same time it must be confessed, that the stories which are told of them by observers deserving of credit, are such as to shake our confidence in spontaneous motion from place to place being a positive test of animal or vegetable nature.

"Among confervæ in ditches are often found little microscopic fragments of organized bodies, some resembling ribands, and separating into numberless narrow transverse portions, others dividing partially at their ramifications, but adhering at their angles like chains of square transparent cases. These are disjointed algæ. When combined, they are motionless, with all the appearance of confervæ, and their joints are filled with the green reproductive matter of such plants; but when they disarticulate, their separate portions have a distinct sliding or starting motion."

Reflecting on these strange plants, it would appear, we think, that though the line of division between the *higher* animals and vegetables is precise and definite, yet that certain of the *lower* animals and certain forms of vegetation closely approximate to each other.

Here we may advert to that strange class of plants termed *Fungi*, under which title botanists comprehend not only mushrooms, toadstools, puff-balls, and similar productions, but also a large number of microscopic plants, forming those appearances which are referred to in popular language by the words *mouldiness*, *mildew*, *smut*, *brand*, *dry-rot*, etc.

Nothing can well be more different than the extremes of development presented by fungi, if we contrast the largest forms with microscopic species; the large fleshy *Boleti*, for example, which grow on the trunks of decaying trees, and the mould-plants, composed of threads too delicate to be distinguished by the naked eye. Investigation, however, has clearly demonstrated that the latter is only a simple phase or type of the former, and that a huge, firm, leathery boletus is merely an enormous aggregation of the fine vegetable tissue constituting a mould-plant or *mucor*. In each is the same plan of development, the same chemical character, and the same mode of propagation. Taking a general view of the fungi, the assemblage of plants included under this term may be described as cellular or filamentous bodies, having a concentric mode of development, and propagating either by means of microscopic granules or seeds, called spores, or by a dissolution of their whole tissue. Contrary to the normal law of plants, they absorb oxygen and exhale carbonic acid.

That these plants consist of a cellular or filamentous tissue may be easily ascertained by means of a microscope of even moderate powers, and one of a higher power demonstrates that the filaments are nothing more than cells drawn out. Sometimes, as in the genus *Uredo*, the cells are spheroidal, having little connexion with each other, each cell containing propagating matter, and all separating from each other in the form of a fine powder when ripe. In plants of a more advanced organization, as the genus *Monilia*, the constituent cells are connected in series which preserve their spherical form, and also contain their own reproductive matter; while in such plants as *Aspergillas*, the cells partly combine into threads forming a stem, and partly preserve their spheroidal form for the fructification. "From adhering in simple series, the structure of fungi advances to a combination of such series into strata, whence result the various kinds of dry-rot, thick leathery substances developing amidst decaying timber; a more complicated structure is thence produced in the form of puff-balls, truffles, and the like, in which a figure approaching that of a sphere is the result. The reproductive cells are indiscriminately confused in the interior of such plants, and the organization is so much complicated, that independently of mere aggregation of tissue we find envelopes of various kinds for the protection of the propagating matter, as in *Argaricus* (mushroom,) and special receptacles for it, as in *Boletus* and numerous others."

"It is probable, however, that in all fungi, and certain that in most of them, the first development of the plant consists in what we here call a filamentous matter, which radiates from the centre formed by the spore, or seeds, and that all the cellular spheroidal appearances are subsequently developed, more especially with a view to the *dispersion* of the species—we say dispersion, not multiplication, because it is certain that the filamentous matter alone is quite as capable of multiplying a fungus as the cellular or spheroidal." As an example in point, we may instance the common mushroom, (*Argaricus campestris*,) the filamentous matter of which is commonly sold, and used by gardeners, under the name of *spawn*, for the artificial multiplication of the species in gardens, etc. The spawn of the mushroom is its stem; the mushroom itself is the fructification of the plant. Here we revert to a point relative to the nature of these singular forms ot vegetable organization, (a point to which we have already adverted,) namely, this, that fungi, unlike other plants, instead of purifying the air by robbing it of its carbonic acid and restoring an equivalent of oxygen, vitiate it by exhaling carbonic acid and absorbing oxygen. This is a singular fact, and goes far to break down the line of demarcation between plants and animals.

There are broad lines of distinction between the great outstanding forms of animal and vegetable life. When we come, however, to those more recondite forms, to those, for instance, which we, perhaps erroneously, call the lowest in organic structure, we find ourselves perplexed; for on the side of the vegetable kingdom we discover among certain microscopic plants, propagation by self-division and locomotiveness, nay, more than this, we find fungi of various kinds exhaling, like the lungs of a quadruped, carbonic acid and absorbing oxygen. Wherein, then, do plants like the confervæ and fungi differ from those zoophytes which are placed within the pale of the animal kingdom, seeing that in functional powers they agree, or nearly so, and that chemistry demonstrates that a fungus is almost like an animal in its intimate constituents and mode of respiration? There are not wanting those who conceive the fungi to be intermediate forms between the true vegetable and the true animal kingdom. This we know, that as soon as the first signs of decay in the mushroom or fungus manifest themselves, or even before, the larvæ of insects burrow through the substance of the plant, and feed as do the larvæ of the flesh-fly on putrescent animal fibre.

There is no affinity, be it remembered, between the rosebud that withers under the incursions of a small caterpillar, and the overgrown mushroom, living and propagating, and yet containing within its disc hordes of minute larvæ, which feed upon its cellular tissue—a tissue, too, that is semi-animalized.

Who, then, shall divide between plants and animals? If we take the oak and the horse, the fir tree and the stork, the reed of the Nile and the crocodile, the distinction is palpable. But when we come to zoophytes, and sponges, and pass from them to confervæ and fungi, who will declare and decide the bounds of their separation? Fungi, like animals, absorb oxygen and exhale carbonic acid gas. As for truly sentient principles, the zoologist looks not for such in the animal kingdom below a certain grade. Wherein, then, we would ask, does the tree, exhibiting in each leaf a centre of vitality, differ from the compound zoophyte, (say *Thuiaria*,) each polype of which is in itself a centre of vitality? It may be urged that in these cases chemical composition is different, and that in the one instance carbon is drawn in and made part and parcel of the system, that oxygen is returned to the atmosphere, and that in the case of the zoophyte, the contrary takes place. Now as regards the zoophytes (we here exclude the higher orders of animals) this has never been proved; while certain plants, as the fungi, are so animalized in structure, as to absorb oxygen, and exhale carbonic acid gas— what more do the lungs of man effect, or of any other air-inspiring animal? We may here also observe, that the elementary body termed iodine is equally obtained from marine molluscous animals and zoophytes, and also from marine seaweeds. We only mention this coincidence, but we do not adduce it as a proof of anything like a coalescence between the animal and vegetable kingdoms. At the same time we consider it interesting, as one among the countless curiosities of nature.

Difficult, however, if not impracticable, as it is for man, even when aided by all the appliances of science, to define the point at which the distinctions between the animal and the vegetable kingdom terminate, it well becomes the reader to pause, and admire the exquisite skill and wisdom with which the great Author of nature has bound together in one vast chain the different orders of created beings. All are mutually dependent. No part of the vast series, however apparently insignificant, could be altered or removed without injury to the remaining portions. The humblest zoophyte has its allotted place in the great kingdom of nature, and in its limited sphere shows forth the glory of its great Creator. Man alone stands an anomaly in the midst of creation. While in an unregenerate condition he fails to discharge aright the high ends of his being. In this respect the meanest insect is his superior, for it perfectly fulfils the object of its existence, while he is false to his. To man was given the glorious privilege of holding communion with his Creator, and of knowing, loving, and obeying him. Sin, however, has separated between him and God. He has broken the holy law of his Creator, and exposed himself to the awful penalty of everlasting misery. How unspeakably important it is, that while there is time he should avail himself of the provision which the gospel has made for his rescue, and seek pardon for his sins through repentance, and faith in the blood of Christ, imploring in fervent prayer the renewing influences of the Holy Spirit, through which alone his nature can again be placed in harmony with the will of God, and made to fulfil the great and glorious objects for which it was originally created!

CHAPTER IV.

REPOSE, OR SLEEP.

AMONG the many points which serve to draw a wide line of separation between organic beings and inorganic matter, there are some which, from the nature of our subject, claim our special notice. And first, we would revert to that peculiar state of repose necessary for the refreshment of the living being, exhausted by thought, by toil, by bodily exertion, which we term sleep. True sleep, if we use the word in a restricted sense, is perhaps peculiar only to mammalia and birds; nevertheless, the lower animals, as reptiles, fishes, insects, and probably others lower in the scale of creation, seek for repose, and pass into a temporary condition of partial lethargy, or deadness to things around them. In some, this condition is assumed on the approach of night; others are awake during the night, and pass the day in slumber.

If we extend the meaning of the term sleep, then must it be allowed that plants, destitute as they are of nerves, or of any of the organs of the senses, undergo a peculiar state of repose—in other words, they sleep.

This condition is very different from that quiescence which takes place after disturbance of the ultimate molecules of inorganic matter, according to the laws of chemical attraction and repulsion; nor oan we compare it to the stillness of matter, after the cessation of physical agitation; although in figurative or poetical language, we say of the ocean, or the lake unruffled by the wind or the current, that it is asleep.

By the sleep of plants we do not mean that change which takes place in so many on the approach of winter, when the sap ceases to circulate, and the foliage falls and withers; this state appears to us to belong to another condition, termed hybernation, experienced by many animals as a means of preserving life, when all supplies of food fail. We allude to a series of phenomena exhibited by plants when in their full vigour, proving, apparently, that they undergo periodical repose, generally during the night; although, as in the case of certain animals, some are awake only during the hours of darkness. It is well known that the blossoms and leaves of many plants become folded up as evening closes; in the tamarind tree, and in some leguminous plants with pinnated leaves, natives of Egypt, similar phenomena present themselves. It was not, however, until Linnæus called attention to the subject that any definite observations had been made relative to the phenomena in question. Accident led him to a train of experiments. He had sown some lotus seeds, and watched with interest the progress of the plants. One of these plants developed two flowers. On visiting them in the evening, to his surprise the flowers had apparently vanished, and as his attempt to discover them was in vain, he concluded that some one had plucked them. On returning in the morning, the two flowers had again made their appearance, but in the evening he found that they had again vanished. He now examined the plants with care, and found the flowers concealed by the leaflets, which, on the approach of darkness, had become attracted to each other, and enshrouded the flowers from view. Struck with this circumstance, he took a lantern in his hand and visited his flower-beds; what was his astonishment to find them no longer with expanded petals! A change had passed over them; they were folded up; and thus he made the discovery of the *sleep of plants.*

As a general rule, it may be stated that on the approach of night flowers close; the leaves of plants become more erect, fold themselves together, and lose in a great measure their vital irritability—a change which is very marked in the leaves of the sensitive plant, the irritability of which retires from the leaf to the petiole or leaf-stalk. At this time, that is during the hours of darkness, a very remarkable change takes place in the functions of plants; during the day they absorb carbon from the atmosphere, and exhale oxygen, but during the night they absorb oxygen and exhale carbon.

Do plants, it may be asked, sleep in consequence of the withdrawal of the light, or is this phenomenon more immediately connected with their organism, that is, with some condition which exists irrespective of the loss of the stimulus of the light to which they are subjected during the day? According to the experiments of Decandolle, by producing artificial day and night, the natural periods of their sleeping and waking may be reversed, and hence, at first sight, we might be led to consider the withdrawal of light as the sole cause of sleep. But, then, how is the functional change to which we have alluded to be accounted for? And besides, as we have said, some are nocturnal, sleeping during the day, and awake during the night, as anagallis, cereus, etc.; moreover, it has been ascertained that flowers, or the leaves of plants, altogether excluded from the stimulus of light, have their regular intervals of opening and folding. Again, what definite end is to be accomplished by this periodical repose? To man and the higher animals, sleep is "nature's sweet restorer;" but can it be said that it is so to plants? It is easier to ask such questions than to answer them.

Let us pass, then, from a consideration of the *sleep of plants*, to a few observations on the *sleep of animals*, more particularly of the higher orders.

Sleep in *man* and the *higher classes* consists in the periodical repose of the organs of the senses, and the greater number of the intellectual faculties and voluntary movements; while the organic processes (namely, the action of the heart, digestion, the peristaltic action of the abdominal viscera, and various secretions) continue. Nevertheless, although the above may be taken as a general definition of sleep, it is not strictly accurate; for in some persons, during health, and in most, during any trifling or more serious disturbance of the system, the mind will be occupied in dreams, in which a long train of occurrences seem to take place, varied according to some previous mental impression, the condition of the nerves, the state of the stomach, or the nature of the transactions in which the person dreaming has actually been engaged. In these dreams, we walk, we run, we ride, we converse, we sing or listen to music, we engage in conflicts, or are in dread of danger. Often these dreams, the acting out of which appears to be so long, are but momentary, and produced by some impulse upon one or more of the organs of the senses; for not unfrequently real sensations are felt during sleep which do not awaken the sleeper, but give rise to a dream, in which he is either a visionary sufferer or actor. A sudden sound, as we have experienced, may be the cause of a dream, the conclusion of which is wound up by an imaginary clap of thunder, or the firing of cannon. Pain suffered during sleep, yet not to such a degree as to rouse the patient, will throw the sleeper into an imaginary position, giving him a false reason for the sensation endured. Dr.

Gregory, for example, who had been recently reading an account of Hudson's Bay, dreamed one night that he spent a winter in that part of the world, and suffered intensely from frost. Upon awaking, he discovered that he had thrown off his bedclothes during sleep; the same gentleman, having applied a bottle of hot water to his feet one night, dreamed that he was walking up Mount Ætna, and felt the ground warm beneath him. "Dr. Reid," says Dr. Abercrombie, "relates of himself that the dressing applied after a blister on his head having become ruffled so as to produce considerable uneasiness, he dreamed of falling into the hands of savages, and being scalped by them."

The horrid visions of nightmare result from the dire oppression of dyspepsia; but the sufferer dreams that some monster of great weight is seated upon his chest or pit of the stomach. A sudden noise, occasioned perhaps by something falling in the room, may originate a dream terminating in an explosion of cannon. We have conversed in whispers with a person sound asleep, but who would, while in that condition, talk about the transactions of the day or week, not unfrequently revealing secrets. But Dr. Abercrombie gives a far more curious example of this kind of *mental wakefulness* during sleep, than any which has ever come under our own notice. He vouches for its truth. The subject was an officer in the expedition to Louisburgh, in 1758. He had this peculiarity in so remarkable a degree, that his companions in the transport were in the constant habit of amusing themselves at his expense. They could produce in him any kind of dream, by whispering in his ear, especially if this were done by a voice with which he was familiar. At one time they conducted him through the whole progress of a quarrel which ended in a duel, and when the parties were supposed to be met, a pistol was put into his hand, which he fired, and was awakened by the report. On another occasion, they found him asleep on the top of a locker, in the cabin, when they made him believe he had fallen overboard, and exhorted him to save himself by swimming. He immediately imitated all the motions of swimming. They then told him that a shark was pursuing him, and entreated him to dive for his life. He instantly did so, with such force as to throw himself entirely from the locker upon the cabin floor, by which he was much bruised, and awoke of course. A remarkable circumstance in this case was that, after these experiments, he had no distinct recollection of his dreams, but only a confused feeling of oppression or fatigue, and used to tell his friends that he was sure they had been playing some trick upon him. Here we have wakefulness of mind, voluntary muscular motion, a scene acted out, and forgetfulness of the whole on being roused to perfect consciousness. Contrast this with the following dream of a long series of transactions which flashed *instan-*

taneously through the mind, but which were yet remembered. "A gentleman dreamed that he had enlisted as a soldier, joined his regiment, deserted, was apprehended, tried, condemned to be shot, and at last led out for execution. After all the usual preparations, a gun was fired, he awoke with the report, and found that a noise in an adjoining room had both produced the dream and awakened him." Dr. Abercrombie relates this instance. Thus, then, the measure of time in dreams does not coincide with that adopted by the mind in a waking state.

It would appear that, of the external senses, touch is the most excitable during sleep, and next to it hearing. Sight is far less excitable, still less so taste, and smell again still less. An unpleasant taste in the mouth, accompanied as it generally is by a deranged state of the stomach, will induce dreams of viands, the imaginary sight of which produces loathing or nausea. In a state of fever, when the mouth is parched, visions of cooling drinks, or grateful fruits, which place the patient in the position of Tantalus, by being ever beyond his reach, are very commonly experienced.

From these remarks, then, by which it seems plain that the periodical repose of the organs of the senses, of the intellectual faculties, and of voluntary motion, is not complete, and that it differs in profoundness in different persons, and in the same person under different circumstances. Sometimes, indeed, during dreams, the faculties of memory and of imagination, and the powers of mental association, appear as if excited to an astonishing degree; and hence arise those dreams which seem almost of a miraculous nature, but of which the truth cannot be doubted. These classes of dreams comprehend warnings, the communication of information relative to bygone events, the revealing of murder, the burial-place of the murdered, and much more. It does not enter into our present plan to analyse the metaphysical subtleties and difficulties which are necessarily involved in a disquisition upon such visions. We refer our inquiring reader to Dr. Abercrombie's "Inquiries concerning the Intellectual Powers," and to other works.* We have here to look at sleep in reference only to organic beings. Somnambulism, however, may require some brief notice at our hands. Somnambulism does not appear, we think, to differ from dreaming so much in its nature as it does in the intensity of the affection of the organic functions. In fact, the somnambulist generally begins by talking in his sleep; the next step in the progress of the affection is that of rising and dressing, or of departing from the room without dressing, often by a dangerous and unusual mode of exit. Some object is to be accomplished; either something is to be done out of the ordinary routine of affairs, some accustomed task is to be performed, or some

* See this subject referred to in the volume of the Monthly Series, entitled Remarkable Delusions.

mental exercise is to be carried on. In the mean time, the actor in this strange scene pursues his course more like some automaton than a human being; his eyes are open, obstacles are seen and overcome, all needful articles are taken from their place, used, and restored; but those who watch the movements and operations of the somnambulist are either not seen, or if imaged on the retina, are not recognised by the eye of the mind. If undisturbed, the patient, his work having been accomplished, returns to bed, reflects upon the past as a dream, and is thunderstruck on finding the operations of that dream real performances. A sudden shock will break the spell, but instant consternation, and mental inquietude afterwards, often result from such a mode of procedure. Sometimes actions are done by the somnambulist which at once bring self-consciousness. A servant girl has been known to rise, descend into the kitchen or wash-house, take off her night-dress, wash it in cold water, even during winter, and immediately put it on again. The shock always roused her. A state very analogous to that termed somnambulism occasionally occurs in the daytime, in which there is the same insensibility to external impressions. The mind is abstracted. Often the memory becomes preternaturally vivid; mental taste appears to have suddenly acquired a new impulse and a new direction; strains of music are sung with singular correctness and sweetness, or a language, unfamiliar as a medium of conversation, is spoken with fluency; circumstances are related; conversations are carried on with imaginary beings; poetry is recited, and, in fact, mental exercises performed by persons, often servants or uneducated individuals, utterly incapable, when awake, of displaying such accomplishments. Dr. Abercrombie adduces a number of most surprising cases, well authenticated, some of which came under his own medical treatment. Epilepsy not unfrequently supervenes on this state of nervous derangement. Nightmare and somnambulism are, perhaps, only phases of the same disease, or may be associated together.

During sleep, the pulse and respiration become naturally slower than during the hours of activity, and the nervous energy and temperature of the body are diminished. Quadrupeds and birds assume such positions as tend to prevent any great decrease of animal heat. Animals crouch, fold their limbs, or double up the body, and those of gregarious habits nestle together, and crowd into a mass. Birds puff out their plumage, and shroud the head beneath the wing. To these rules there are exceptions; for some quadrupeds, as the horse, as often sleep standing as lying down; and some birds sleep standing on one leg, curiously balancing themselves—a proof that, however profound sleep may be, the *muscular sense* remains unaffected, the sensibility to the position of the limbs remaining undiminished. This fact is, indeed, one of the many proofs which sir Charles Bell adduces of our possession of this sense. "We awake with a knowledge of the position of our limbs; and this cannot be from a recollection of the action which placed them where they were; it must, therefore, be a consciousness of their present condition. When a person in these circumstances moves he has a determined object, and he must be conscious of a previous condition before he can desire to change or direct a movement." After long-continued fatigue, sleep overcomes the frame, often in unnatural attitudes, or under unwonted circumstances. Horse-soldiers and couriers have been known to sleep for hours while travelling on horseback, sustained in their saddle by the operation of the muscular sense alone. Nay, foot soldiers have slept as they have wearily marched along, and violin players have been seen to continue playing while wrapped in the slumber of fatigue. Slumber comes upon us insensibly, indistinctness steals over the senses, our eyelids close, hearing fails, the limbs relax, we instinctively assume an easy position, and sleep wraps us in a mantle of temporary forgetfulness. Yet, as we have said, the muscular sense is not lost. Hence the "wet sea-boy," perched aloft amidst the shrouds, slumbers in security.

But there is a point of fatigue, both bodily and mental, which produces such a condition or the nervous system as to prevent sleep, and induce a morbid watchful restlessness—a restlessness which accompanies many diseases, rendering the couch so uneasy that the sufferer exclaims: "Would God it were even!" and at even, "Would God it were morning!" Deut. xxviii. 67. On the other hand, without that bodily fatigue which renders "sleep sweet to the weary," a period of natural sleep will alternate with a natural period of wakefulness. This at least is observable in the higher animals, among which, in the course of every twenty-four hours, a portion of time varying in its length requires to be devoted to slumber. This period of repose may be broken for a time, but the interruption cannot be carried beyond a certain limit of duration, when sleep will, in spite of every effort to the contrary, assert its authority. The most anxious nurse in the sick-chamber, accustomed to watching as she may be, requires a portion of repose. Even grief at length yields to sleep.

To *cold-blooded animals* sleep at short intervals is, perhaps, less necessary than it is to warm-blooded animals. Their temperature is little above that of the medium they inhabit, and their muscular irritability is extreme. Yet, although fishes cannot exclude the light from their eyes, by means of eyelids, they sleep, and the eye loses its perception of surrounding objects. They choose out some obscure spot for repose, or float on the surface of the water, basking in the beams of the sun. The pike, for example, thus slumbers, and in this state is often hauled upon the bank by means of a running noose of wire, (at the end of a stout rod,) adroitly passed over it. In like manner, snakes slumber on sunny banks, and with caution may be surprised. The same observation applies to the active little lizard of our hedgerows and copses.

In our species the necessity for sleep varies; infants pass the greatest portion of their time in sleep, and sink into repose after the reception of nutriment. In very aged persons, a preponderating duration of sleep over wakefulness seems necessary; though to this rule exceptions are often to be met with. Children and young persons sleep more soundly than persons more advanced in life. There is a tendency to sleep, or at least to rest in quiet after repletion. Animals that chew the cud lie down to ruminate. The dog curls himself up and sleeps after a good meal; and, but that necessity forbids this indulgence in mankind generally, at least among those whose hands are called to labour, we should be apt to imitate our canine companion. Invalids, and persons of feeble delicate frame, require this indulgence, if in their case we can call it indulgence. In hot countries, men and animals take their siesta during the fervid heat of mid-day. Sleep is induced by narcotic medicines, and by extreme cold, which acts as an all-overpowering narcotic, paralysing the energy of the brain and spinal cord, and in no very long time inducing coma, or torpor, and death. In this case, however, as in the results of an over-dose of narcotic drugs, the coma or torpor thereby induced, like the coma of apoplexy, must not be confounded with *true sleep.*

From the very abundance and frequency of God's mercies, we are in danger of forgetting our obligations to thankfulness on account of them. Amidst the commonest but richest of these mercies, healthy and refreshing sleep may fairly be numbered, and to have enjoyed it for a series of years unbroken by disease and pain may well call forth our sentiments of gratitude to the Author and Giver of all good gifts. Sleep is, too, one of those natural arrangements which seems like many others to be fraught with rich spiritual analogies. It has been appropriately termed the image of death, and by its nightly recurrence is well calculated to habituate the pious mind to contemplate the arrival of that period when the eye shall close on the things of time for ever. Happy, indeed, are they who each night, as they close their eyes in slumber, can through a living faith in the all-sufficient atonement of Christ exclaim, "Father, into thy hands I commend my spirit—thou hast redeemed me;" and cheerfully refer it to an almighty Disposer, whether they awake in this world or another.

CHAPTER V.

HYBERNATION.

Besides the ordinary oft-recurring sleep, so essential to mental refreshment, and to the harmonious performance of our animal functions, many mammalia, and all reptiles, undergo annually a long duration of sleep, extending for many weeks, or even months, and which is necessary to their very existence. We allude to that winter sleep, termed Hybernation, from *hyberno*, to take up winter-quarters.

Hybernation is by some called *Torpor*, but we may state at once, that this *torpor*, if we admit the use of the word, is not that kind of torpor which results from the influence of excessive cold. Torpidity from cold is a morbid condition—it is, in fact, a suspension of animation, a reduction of the vital energy, continuing until death closes the conflict between the vital principle and the overpowering narcotic. But the long sleep which we call hybernation, is intended as a preservative of life; extreme cold will destroy an hybernating animal; at the same time, a moderately low temperature conduces to hybernation, which as in the case of ordinary sleep comes on gradually, with diminished activity, and a diminished sense of hunger. When a hybernating animal feels the drowsiness which ushers in its winter sleep, and is under the influential agency of a certain degree of cold, it is urged by instinct to make preparations of self-defence against an excess of severity, the effects of which would be fatal. It therefore retires to its hybernaculum, or winter dormitory, closes itself in, falls into a tranquil sleep, which in due time passes away, and is succeeded by a renewal of activity.

The intensity or profoundness of this sleep varies in different animals; in all, the pulsations of the heart become feeble and slow, the respiration is also very slow, and frequently intermits for a considerable interval, or even becomes altogether imperceptible. The temperature of the body sinks greatly, and in some instances is but little elevated above that of the surrounding medium, perhaps only by two or three degrees. In very profound hybernation, the body feels cold and stiff, the senses are in a state of oblivion, and nervous sensibility is greatly reduced, insomuch that severe wounds, and even electric shocks, are insufficient to awaken the lethargic sleeper to consciousness. During this sleep, a diminished weight of the body takes place, to be restored by food after reviviscence.

Confining ourselves for the present to *quad-*

rupeds, we may observe that different species are led by unerring instinct to choose different kinds of winter dormitories, and make different arrangements. The marmot excavates a deep burrow, in which it makes a bed of dried grass and moss, to which it retires in autumn. Among burrowing quadrupeds, the hamster is remarkable for the extent of its underground galleries and chambers, which constitute its summer abode and winter asylum. The dormouse makes a nest in the crevice of a tree, or in the thickest part of a dense brake, to which it retreats, coiling itself up into a ball, and thus in its snug dome-covered dormitory, made of moss and herbage, awaits the approach of sleep. The hedgehog makes a warm soft nest of moss and leaves, under the root of some old tree, in the hole of a bank, under the covert of logs or masses of timber, and there passes the winter in profound repose. The mole descends deeper into the soil in winter, but does not hybernate; on the return of spring it leaves its deep fortress, and commences its labours on the surface of the earth. Whether the shrewmouse hybernates, in the strictest sense of the word, we do not well know; most probably it does so to a certain extent during the inclemency of mid-winter. Bats hybernate in the hollows of trees, in old ruins, in church-towers, in barns, in caves, and similar localities; they make no nest, but hang suspended by the hinder claws. In every case, however, the aim seems to be to secure a shelter from extreme cold, and the maintenance of a degree of temperature conducing to a peculiar condition of the system, without involving the loss of the vital principle.

Most mammalia hybernate in solitude ; some, however, pass the winter in company. Families of marmots and of hamsters associate in one chamber. Bats generally hybernate in clusters huddled together in their dark recess, and thereby, perhaps, maintain an atmosphere around them somewhat higher than that of the wintry wind, as it sweeps over the open country. Hybernating animals are always sheltered from the influence of cutting winds, which reduce the bodily temperature, and tend to the destruction of life, much more rapidly than would be effected by the simple depression of atmospheric temperature, the air being in a state of quiescence. In the former case, the abstraction of caloric from the body takes place with fearful rapidity. There are two conditions of hybernation, namely, *perfect* and *imperfect*, to which we may here allude ; but there is a third kind of hybernation, giving full latitude to the word, unconnected with sleep, and yet necessary for the preservation of life.

In our country the insectivorous hedgehog affords a familiar specimen of *perfect* torpidity —the dormouse of *imperfect* torpidity. "The hybernation of the hedgehog," says Mr. Bell, "is perhaps as complete as that of any animal inhabiting this country, and much more so than that of many of the *Rodentia*, (dormouse, long-tailed field-mouse, etc.,) which retire indeed to winter retreats, but awaken at intervals to eat of their treasured hoard of nuts or grain, when called into temporary life by a day of unwonted mildness. The hedgehog, on the contrary, lays up no store for the winter, but retires to its soft nest of moss and leaves, and rolling itself up into a compact ball, passes the dreary season in a state of dreamless slumber, undisturbed by the violence of the tempest, and only rendered still more torpid by the bitterest frost."

The marmot lays up some store of provision for consumption in early spring when its trance is over, and ere yet the mountain pastures afford its food. Its hybernation, however, is perfect. About the middle of September, these social animals retreat to their chambers underground, closing the entrance passages with earth, dried grass, etc.; they then very soon fall asleep, and so continue till the beginning of April. On their retirement, they are fat and in good condition—they awake lean and meagre, and need the restorative afforded by their little store of grass or herbage before they venture forth to crop the new verdure of the high alpine slopes, where winter lingers and the snow long remains.

In the case of the bat, insectivorous as it is, hybernation is imperfect. A little increase of temperature rouses it into activity, nor is it then at any loss for food, for a warm winter's day awakens also from their hybernation myriads of gnats and minute insects which afford a passing meal; it then retires to its recess, suspends itself, and falls asleep again. Some species, doubtless, hybernate more perfectly and enduringly than others. We have seen the Pipistrelle bat all on the alert on the afternoon of a fine day in December, in February and March; in short, as Mr. Bell observes, "this, the most common of our indigenous bats, will sometimes make its appearance in fine weather in almost every month of the year," not restricting itself, we may add, to the obscurity of evening.

The great bat or noctule (*Vespertilio noctula*) appears, however, to sleep very profoundly. It is never seen abroad in winter; it appears in April, and retires at the close of July or in the month of August; we have ourselves seen it flying high over a grove of sycamore trees in August. It hybernates in considerable assemblages. Moths and large coleopterous insects, as the chaffer-beetle, appear to constitute its food.

Another example of imperfect or partial hybernation is afforded us by the store-collecting hamster, that scourge of the corn-plains of middle and eastern Europe. In this case, the duration of *true hybernating slumber* is very short, and does not take place until long after the animal has retreated to its subterranean chamber.

We have sufficiently indicated what we mean by perfect or *continuous*, and imperfect or *inter-*

rupted hybernation, to which latter, *brief* hybernation belongs rather than to the former. We may now, therefore, advert to that third kind of hybernation previously alluded to, during which sleep beyond what all quadrupeds ordinarily require, is either not at all experienced, or only in a degree so trifling as not to have attracted especial notice. The *beaver* shall be first cited as affording an example in point. The beaver is a gregarious animal of semi-aquatic habits. During the summer, the beavers composing a colony roam about at pleasure, and it is during that season that they fell the wood necessary for repairing their houses and dams, or for building others, commencing the latter about the end of August. The use of the dam is to secure a great depth of water beyond the freezing power to congeal at the bottom ; where the water is of sufficient depth a dam is not constructed. The houses vary in size and in the number of lodging-rooms. These rooms do not communicate with each other, but each has a separate entrance under the water. Like the dam, the houses are constructed of wood, stones, and mud, and during the winter the compact mass becomes frozen as hard as stone. When complete, the houses have a dome-like figure, with walls several feet thick based upon the bed of the river, and emerging from four to six feet above the water. The only entrance is deep under water, below a projection called by hunters the angle, and consequently the water fills a great part of the vault itself. Near the entrance, and also deep under water, is laid up their winter store, a mass of the branches of willows and other trees, on the bark of which they feed. These they stack up, sinking each layer by means of mud and stones, and often accumulate more than a cartload of materials. Within each house is a dry portion of flooring above the level of the water; on this the animal rests and enjoys its food, brought in as wanted from the submerged magazine. Here, then, we have an animal which forms for itself a solid winter habitation, which lays up a store of food in abundance, which retreats to its house on the approach of cold season, and there resides, patiently awaiting the return of spring, and secure from every foe excepting man. Thus the beaver may be said to hybernate, although it continues active, without sinking at any time into lethargy. Nor is it so much a prisoner as might be at first supposed, for besides its house, it makes burrows or excavations in the bank of the river, the entrance being deep under water, below the freezing line, and the passage inclining upwards, until it terminates in a dry cell or chamber. This is a place of refuge. Where beavers are scattered, and lead a solitary life, as occurs sometimes in America, where colonies of them are almost entirely exterminated by the hunter, the animals do not attempt to build houses or dams, but permanently reside in such burrows. It is thus that the scattered beavers along the Rhone exist, living singly or in pairs

As another example of this mode of hybernation we may adduce the *musquash*, or musk-rat, (*Ondatra zibethica*,) so called from its musky and rather fragrant odour. This animal inhabits the higher latitudes of North America; small grassy lakes or swamps, or the grassy borders of slow streams are its favourite haunts. Vegetable matters are its principal food, as roots, tender shoots, the leaves of various *carices*, etc., to which it adds fresh-water mussels. Like the beaver, it is semi-aquatic, swimming and diving with the greatest facility. Like the beaver, too, it builds winter habitations, but of far less solidity and durability than those of the former. Its fur is an object of commerce, hence thousands are annually slaughtered by the hunter. Between four and five hundred thousand skins are annually imported into Great Britain. Dr. Richardson thus describes the house of the musquash:— "In the autumn, before the shallow lakes and swamps freeze over, the musquash builds its house of mud, giving it a conical form, and a sufficient base to raise the chamber above water. The chosen spot is generally amongst long grass, which is incorporated with the walls of the house from the mud being deposited amongst it; but the animal does not appear to make any kind of composition or mortar by tempering the mud and grass together. There is, however, a dry bed of grass deposited in the chamber. The entrance is under water. When the ice forms over the surface of the swamp, the musquash makes breathing-holes through it, and protects them from the frost by a covering of mud. In severe winters, however, these holes freeze up in spite of their coverings, and many of the animals die. It is to be remembered, that the small grassy lakes selected by the musquash for its residence are never so firmly frozen nor covered with thick ice as deeper and clearer water." Here, then, we have another example of an animal hybernating while it continues in full activity.

The following little poem "On the Musquash, or Musk-rat," by an anonymous writer, is very descriptive, and is so little known that we may be excused for quoting it. It is most probably the production of an American author, certainly of one practically conversant with the habits of the animal:—

"Where the wild stream, half-chok'd with sedgy weeds,
Winds its dark course through transatlantic meads,
And, creeping onwards, joins the river's flow
That tumbles down in swift cascades below,
Bound for St. Lawrence and his islets,—there
Inhabit many a happy musk-rat pair,
That rove the verdant shores, and pluck the weed,
And in fond concert on the foliage feed;
Or gather fruits, or dive where in its shell
The pearly mussel, and green mya dwell—
Sometimes their food;—or stray delighted where
Spreads o'er the bank the strawberry's wild *parterre.*
Here on the bank the mother finds some cave
To nurse her young, beside the silent wave.
And all are foragers. Soon as her brood
In ripen'd strength may learn to seek their food,
Then oft at midnight, by the moon's pale beam,
Their waving shadows flit beside the stream,
And vanish quick; whilst sweet as vernal hay
Their fragrance breathes where'er the ramblers stray.
But when the fading leaves of autumn fall,
Their guardian gnomes the scatter'd wanderers call,

And teach their bands in fed'ral strength to form,
Ere winter comes, a shelter from the storm.
The solid structure fram'd with twisted reeds,
Plaster'd with mud, and interlac'd with weeds
Four cubits measures in its space around,
Rais'd like a little turret from the ground.
Within, thick buttress-steps around supply
Strength to the walls, and keep the lodgings dry;
At top, a rounded cupola or dome,
Twelve inches thick, roofs in this winter home.
Here, with their young, whole families repose,
Whilst gather'd o'er them rest the winter snows.
Yet do they not, like marmots, hoard and sleep,
But wander still, and forage in the deep,—
Like mining moles, through hollow pathways stray
To spreading roots, or catch retiring prey;
And still beneath the frozen stream they feed
Upon the water-lily and the reed.
And thus they live, excluded from the light,
In total darkness—in perpetual night.
At length, the sun resumes, as winter yields,
A strengthening empire o'er the wither'd fields;
The ice dissolves, the snows all melt away,
And leave expos'd the musk-rat's house of clay.
Then comes the hunter, and his efforts tear
The dome-roof off. Then pours the day's full glare
Upon the darkness, and bewilders all,
While in their house the easy victims fall.
For e'en their gnomes the sudden burst of day
Frights from their post, and drives confus'd away.
But soon they rally, part their lives redeem,
And through their galleries hurry to the stream.
And these again are wanderers as before,
Within the river, and along the shore."

We are not aware that any writer has considered the winter retirement of animals for protection against the cold, without entering into a state of torpidity, (the beaver and the musquash for example,) as coming within the pale of hybernation. A little consideration, however, will serve to place the matter in its true light; for, from *continuous* or perfect hybernation, we are led through different phases and degrees of *interrupted* or partial hybernation, to *sleepless* hybernation. What shall we say of the hybernation of the common brown bear of northern Europe? In this case is torpidity continuous or interrupted, or is the hybernation sleepless? We take the following passages from our own writing:—"The general habits of the *bear* are well known; unsocial and solitary, these animals frequent the gloomiest recesses among the mountains, glens, and caverns, and the depths of forests; there they dig or enlarge a cave in which to dwell, or usurp the hollow of some huge decayed tree, or form a sort of rude den under the covert of a maze of intertwined branches, lining their habitation with moss. Here they pass the winter in a *state bordering* on torpidity; and it is during this *retirement*, in January, *that the female brings forth her young*. . . . When the bear retires to winter quarters, on the approach of the cold season, it is very fat; but on coming forth in the spring it is generally observed to be very lean, the fat having been absorbed for the nutriment of the system during the animal's torpidity. But a query here suggests itself—Is the female who produces her young, and has to attend to them, torpid? And can she suckle them without receiving any aliment herself? This is very improbable, and tends to prove that the seclusion of the animal is neither so absolute, nor its seclusion so complete as is generally asserted. That bears support themselves in their winter retirement by sucking their paws is a vulgar error." Error, however, as this is, it shows the prevailing idea among the peasantry of the north, that the bear does not sleep like the dormouse or the marmot. It is somnolent, but not torpid, nor is there any muscular rigidity. The following passage occurs in the History of the American Black Bear:—"In the fur countries this species usually hybernates, selecting a spot under a fallen tree, where it scratches a hollow in the earth. Here it retires at the commencement of a snow-storm, and the snow soon furnishes it with a close, warm covering. Its breath makes a small opening in the snow, and the quantity of hoar-frost which occasionally gathers round the opening serves to betray its retreat to the hunter. In more southern districts, where the trees are larger, bears often shelter themselves in the hollow trunks. It has been observed by the Indians, that unless bears are very fat on the approach of winter they do not hybernate; and as the males are often thin and exhausted in September, should the winter set in before they have time to recover their fat, they migrate southwards in search of food. So carefully do the females with young conceal themselves, that Dr. Richardson's numerous inquiries among the Indians of Hudson's Bay ended in the discovery of only one man who had killed a pregnant bear." We next refer to the account of the Polar Bear:—"It is stated, on the best authorities, that the *male does not* hybernate, but that the female, on the approach of the severer season, retires to some rift among the rocks or ice, or digs a lair in the frozen snow; the falling snow drifts over the den, covering it to a great depth, a small aperture for breathing being always open. In this retreat, about the latter part of December, she brings forth two cubs, and in March quits the den with them, then about as large as a shepherd's dog, and prowls abroad, lean, gaunt, and ferocious; hunger, and the presence of her offspring, adding fury to her savage temper. The male wanders about the marshes and adjacent parts until November; he then goes out to sea in quest of seals, and becomes very fat. It often happens that he becomes drifted out from the coast on a floating field of ice."

In these accounts of three species of bears we are presented with different modifications of hybernation. First, the common brown bear hybernates, but whether in a state of torpidity or not is very doubtful; the female cannot become torpid, and the reason is obvious. Secondly, the American black bear does not invariably hybernate; their males wander southward, but the females seek a winter retreat, not for the sake of slumber, but as a nursing den. Thirdly, the male of the Polar bear *roams about* in quest of prey during winter; while the female, in her snow-covered seclusion, brings forth and nourishes her young.

Thus, then, there are grades in the state of hybernation, from continuous torpidity to partial torpidity, and thence to wakeful activity. Then we come to accidental seclusion, with or without torpidity; or to seclusion without torpidity in the female; but with more or less torpidity, or, on the contrary, with full activity, without seclusion, in the male.

Man himself hybernates. In the high Polar regions he retires to his winter hut, suspends all out-of-door employments, hangs up his fishing-spears, closes the apertures of his hut or den, lights his lamps of train oil, clothes himself in a furry vestment, feeds upon the coarse (and to us disgusting) produce of the summer's fishery or chase, and, unexcited by the inspiration of intellect, slumbers away the greater portion of the dreary season, till roused by the advance of spring, he issues forth to engage in necessary toil, and encounter inevitable danger.

But we must here revert to the condition of those animals which hybernate in decidedly torpid quiescence.

We have said that extreme cold, in the case of hybernating animals, does not produce torpidity, but death. Torpidity is latent life—death is the extinction of life. Torpidity may continue, as in the seeds of plants, or in animalcules, dried up for an unknown number of years, but death has not taken place. Thus, then, the torpid hedgehog or dormouse, dead as it may appear, will revive; but if exposed to extreme cold it dies, and no revivification will take place, for its death is real, not apparent. If we expose an animal, which naturally becomes torpid at a certain season of the year, to excessive cold, and allow it no opportunity of sheltering itself, it will certainly perish. If we subject an animal hybernating in a state of torpidity to excessive cold it will revive, and probably die; and here cold seems to act as an increase of temperature would do. Mangili placed a torpid marmot, which had been kept in a temperature of 45°, in a jar surrounded with ice and muriate of lime, so that the temperature of the jar rapidly sank to 16°. In about half an hour, the respiration of the animal became quickened, and it exhibited symptoms of returning animation. In sixteen hours, it was completely revived, and made many efforts to escape. He also placed a bat under a bell-glass, where the temperature was 29°, and where it had free air. Respiration soon became painful, and it attempted to escape. It then folded its wings, and its head shook with convulsive tremblings. In an hour no other motions were perceptible than those of respiration, which increased in strength and frequency until the fifth hour. From this period the signs of respiration became less distinct, and by the sixth hour the animal was found dead. He also exposed a torpid dormouse, from a temperature of 41° to a cold of 27° produced by a freezing mixture. Respiration increased from ten to thirty-two times in a minute, and without any intervals of repose. There were no symptoms of uneasiness, and the respirations seemed like those in natural sleep. As the *temperature rose*, respiration became *slower*. He then placed it in the sun when it awoke.

Two hours afterwards, having exposed it to the wind, respiration became frequent and painful; it turned its back to the current, without, however, becoming torpid. Here, then, we see the difference between the torpidity of hybernation, and the torpidity resulting from excessive cold. In the former case, the result is a renewal of health and activity—in the latter, death. In the most torpid animal, during hybernation, life is latent—in the animal that perishes from cold, life is extinguished. Why, it may be asked, do not all animals in northern climes hybernate, seeing that many do? There is something, no doubt, in the constitution of some animals which predisposes them to assume the condition described; and we shall find that where such a constitutional predisposition is given, two objects are aimed at,—preservation from a degree of cold which the system cannot bear with impunity, and preservation from starvation in consequence of the failure of natural diet. To many animals this does not apply: winter may thicken their coat, or, as in the case of the ermine and the arctic fox, whiten their close deep fur, but their food is as much within their reach in winter as in summer, and their system is adapted to sustain the utmost severity of the former; their constitution, therefore, is not predisposed to a state of torpid hybernation, such a preservative not being essential. If animals which naturally hybernate in a state of torpidity be kept in an artificially high atmosphere, as in a warm room, and supplied with food, they either do not hybernate at all, or sink into transient fits of torpidity, reviving, and slumbering for a few hours only. We have never known any of the marmots kept in cages at the menagerie of the Zoological Society to become truly torpid during the winter months. But we suspect that this wakefulness, unnatural in the extreme, is kept up at the ultimate expense of life, being a tax upon the system. We have known dormice to be kept in a warm room during the winter, and to continue tolerably lively; but we have observed that none have survived the spring, at which period they would, in a state of nature, have awakened from their trance. Is not, then, this strange suspension of vitality necessary for their well-being, and a wise provision ordained by Him who regardeth even the fall of a sparrow?

Animals in a state of rigid torpidity do not bear with impunity a sudden revival, by exposure to warmth. We have seen bats and dormice immediately die, after one or two such artificial resuscitations. The few animals which we have seen artificially revived appeared to experience painful or oppressive sensations. The breathing was quickened—occasionally deep inspirations were made; the rigidity of the muscles relaxed, the eyes were opened, attempts were made to use the limbs; but the movements were all unsteady, and the creature seemed bewildered and confused. In a state of nature, the transition is gradual; the animal quietly awakes from a natural slumber, and the system suffers no shock. Animals which hybernate at a certain period of the year, in obedience to a protective law, will not hybernate if exposed to cold at another season; and if the cold be intense they will perish, a fact proved by the experiments of Mangili.

CHAPTER VI.

HYBERNATION OF BIRDS—TORPIDITY OF REPTILES.

PASSING from quadrupeds, (or mammalia,) we turn to *birds*. Birds do not hybernate; and all the stories told, and formerly believed, even by Klein, Linnæus, and others, of the submersion of swallows during the winter in lakes or rivers, must be rejected as unworthy of credit. Cuvier, in his *Règne Animal*, vol. i. p. 396, says of the sand martin, (*Hirundo riparia*,) "It appears that it becomes torpid in winter, and that during this season it buries itself at the bottom of the marshes."

We are astonished that Cuvier should have lent the authority of his great name to such a theory, and equally so to read the following passage by baron Humboldt: "The circumstance of the sand martin sometimes burying itself in a morass, is a phenomenon which, while it seems not to admit of doubt, is the more surprising, as in birds respiration is so extremely energetic, that, according to Lavoisier's experiments, two small sparrows in their ordinary state decomposed in the same space of time as much atmospheric air as a porpoise.* The winter sleep of the swallow in question (*Hirundo riparia*) is not supposed to belong to the entire species, but only to have been observed in some individuals."†

Gilbert White, who, with others of his day, firmly believed in the submerged winter state of the swallow tribe generally, in one passage expresses his suspicions that the sand martin, which often makes its appearance at the beginning of April, does not leave its wild haunts at all, but that the flocks are secreted amidst the clefts and caverns of those abrupt cliffs where they usually spend their summers. But he just previously states, that on a diligent examination of their breeding-holes in sand banks nothing but the old nests were found.

Dismissing the submergence theory as utterly untenable, let us pass to another view of the question—Do swallows or any other birds hybernate, either regularly or accidentally, in holes or crevices, like bats? Markwick says, "I have frequently taken notice of all those circumstances which induced Mr. White to suppose that some of the *hirundines* lie torpid during winter. I have seen so late as November, on a finer day than is usual at that season of the year, two or three swallows flying backwards and forwards under a warm hedge, or on the sunny side of some old building; nay, I once saw, on the 8th of December, two martins

* Lavoisier *Mém. de Chimie.*
† Milne Edwards' *Elemens de Zoologie*, 1834.

flying about very briskly, the weather being mild. I had not seen any *considerable* number either of swallows or martins, for a good while before; from whence, then, could these birds come, if not from some hole or cavern where they had laid themselves up for the winter? Surely it will not be asserted that these birds migrate back again from some distant tropical region, merely on the appearance of a fine day or two at this late season of the year. Again very early in the spring, and sometimes immediately after very cold severe weather, on its growing a little warmer, a few of these birds suddenly make their appearance, long before the generality of them are seen. These appearances certainly favour the opinion of their passing the winter in a torpid state, but they do not absolutely prove the fact, for who ever saw them reviving of their own accord from their torpid state, without being first brought to the fire, and as it were forced into life again? *soon after* which revivification they *constantly* die."

Now, no one ever yet saw a swallow in a state of hybernation. It may be that half-starved birds, benumbed with cold, have been found, ere quite dead, in some hole or place of shelter, and that warmth has revived them for a little time, death terminating their suffering. But is this hybernation? Again, in advance before the great cloud of swallows sweeping over Europe from their African winter quarters, on their return to their usual summer haunts, a small number of stragglers, like an irregular vanguard, appear. From among these, if the weather be open, a few urge their way from milder latitudes northwards, and occasionally visit us, or are drifted by the wind to our shores. Should the cold suddenly set in, they wing their way southwards again, and await their proper time; some perhaps perish. So also with the clouds of swallows that leave us in autumn, (end of September,) a few stragglers linger behind, as if to bring up the rear. These are the young of the last brood; and many, as we have seen, remain with us till the middle of October, some few linger till the commencement of November, and perhaps a still fewer number till the beginning of December. If, by reason of their unfitness for a protracted flight, a few linger till the cold of winter surprises them, they creep into some shelter, (perhaps their old nest,) become benumbed, starved for want of food, and perish.

Other migratory birds, it has been asserted, sometimes hybernate. Bewick says, "A few years ago a young cuckoo was found in the thickest part of a close whin-bush. When taken up it presently discovered signs of life, but was quite destitute of feathers. Being kept warm, and carefully fed, it grew and recovered its coat of feathers. In the spring following it made its escape, and in flying across the Tyne it gave its usual call." If there be any truth at all (which we greatly doubt) in this unauthenticated story, how can it be said that a half-starved nestling cuckoo, fallen from some nest in a whin-bush, was comfortably hybernating like a dormouse? "The bishop of Norwich, in his Familiar History of Birds, records an instance of about forty cuckoos being congregated in a garden, in the county of Down, from the 18th to the 22nd of July, and, with the exception of two, which were smaller than the rest, taking their departure at that time. These were, no doubt, all young birds of the year, and it is probable that the two smallest were never able to follow the others, but remained to perish. Of such a character were the benumbed, denuded birds, which have been occasionally found in hollow trees, or the thickest part of furze bushes, whither they had crept for shelter, and which have been noticed by Willoughby, Bewick, and others."

We now pass to *reptiles* and *amphibia.* In the colder and temperate regions of the globe, all terrestrial reptiles, and all amphibia, (as far as the habits of the latter are known,) pass the rigorous season of the year in a state of torpid hybernation. Those who have kept the common European *tortoise* in a garden must have observed how, as autumn advances, the animal excavates a sort of hole, or burrow, in the ground, in an oblique direction, into which it finally retires, and thus buried passes into a state of lethargy. It has been ascertained that during hybernation digestion entirely ceases; but, generally speaking, the tortoise refuses to take food for some weeks previously to retiring to its subterranean dormitory. Yet it has been our lot to dissect tortoises which have died in the spring soon after resuscitation, and before taking food, and we have found the vegetable matters, with which the stomach had been filled during the previous autumn, to be totally unchanged. Professor Bell records a similar case which came under his own cognizance.

Snakes hybernate, and though generally anti-social in their habits, often, as in the case of the common ringed snake, the viper, and the blind worm, associate together in considerable numbers, drawn, as it were, from various quarters to one common place of concealment, (a deep hole under the roots of a tree, a cavity or excavation under stacks of firewood, under the covert of dense thickets, hedges, etc.,) in which, as if to preserve some degree of warmth, they coil and intertwine together. Our British amphibia, as the *frog*, *toad*, *newt*, etc., retire on the approach of winter to their hybernating retreats, and there pass into a state of absolute torpidity. "This is generally in the mud at the bottom of the water, where they are not only preserved in a nearly equal temperature, though at a low degree, but also secured from external injury. Here they congregate in multitudes, embracing each other so closely as to appear almost as one continuous mass. On the return of spring they separate from each other, and merge from their retirement, and recommence their active life." Why, it may be asked, are these frogs not drowned, seeing that respiration is rendered impossible? The reason is evident. In the first place the action of the heart is all but, or indeed altogether, suspended, so that the merest trifle of oxygen, if any at all, is required for the purposes of the system. In the next place, the large cellular lungs most probably contain a sufficient quantity of air for the needful supply; and thirdly, these animals respire by means of the skin as well as by the lungs; not, however, that this process takes place during their hybernation.

Wherein the hybernation of reptiles essentially differs from the hybernation of quadrupeds, except in its extreme degree of torpidity, and the more complete suspension of all the vital functions, we cannot tell. As they are not frozen to death, but revive, surely their torpidity is preservative. It is true that cold-blooded animals are always inert, and some torpid, when the temperature of the air is even moderately low, and that they require warmth. But, if not in so marked a degree, yet warm-blooded animals, according to the latitude for which they are respectively destined, become inert under a certain degree of cold, and require a certain temperature for the fulfilment of the duties of active life. Reptiles, like marmots, retire to their hybernacula; instinct directs them to prepare against the advancing winter; they become torpid, and thus revive. Mere cold, however, will not make a dormouse torpid, but it will make a reptile torpid, and the animal may remain in this condition for a great length of time.*

It does not appear that the marine tortoises or turtles retire to hybernate; they tenant a medium, the temperature of which is generally uniform, and they have the means of migration. They traverse leagues of ocean, in order to resort to their favourite breeding places, and afterwards revisit the expanse of the sea to enjoy existence. Among the amphibia, the *Amphiuma* of Georgia, Carolina, Florida, etc., generally buries itself in the mud of the swamp or marsh in which it habitually resides. With respect to the *Proteus* of the subterranean lakes of Carniola, nothing is known. Here, also, we may refer to those singular animals, the *Lepidosirens*, of which one species is a native of the Amazon, in South America, (*L. Paradoxa;*) a second, (*L. Annectans,*) of the river Gambia, and which seem to link the amphibia and fishes together. Of these, the latter, from Africa, appears at a certain season of the year to become torpid, burying itself in the mud.† From what we can glean respecting the *Gambia Lepidosiren*, it appears to bury itself

* Spallanzani kept frogs, newts, and snakes, in a torpid state in an ice-house, where they remained for three years and a half, and readily revived when again exposed to the influence of a warm atmosphere. No hybernating quadruped could have been so kept; it would have awakened and died. So far there is a difference between the torpidity of reptiles and the hybernation of quadrupeds.

† We refer our reader for details to "A Popular History of Reptiles," published by the Religious Tract Society.

in the mud along the banks of the river it inhabits, and there folding itself up, to pass a great portion of its time in a state of torpidity. Pieces of hard clay, bearing the impressions of these animals, have been brought over to England. Sir W. Jardine, in his observations on this creature, refers to the circumstance as follows :—" Miss Weir, in allowing us to examine the specimens of the fish, accompanied them with a piece of the hard clay alluded to in the Transactions of the Linnæan Society, bearing the impression of the animal, as if it had lain for some time imbedded in it, and with the earth in such a state as to allow the form of the cast to be retained. Fish taken in the summer of 1835 on the shore of Macarthy's Island, about 350 miles up the river Gambia, were found about eighteen inches below the surface of the ground, which during nine months of the year is perfectly dry and hard; the remaining three months it is under water. When dug out of ground and put into water, the fish immediately unfold themselves and commence swimming about. They are dug up with sharp stakes, and used for food." Most probably the Amazon species undergoes a similar torpidity. This long torpidity of nine months, during the dry season of the year, when the ground is parched, and the river low, will come under the head of what we denominate æstivation, and not hybernation; of this, however, we shall speak hereafter.

If the lepidosiren comes within the pale of the fishes, it is one among the few of that class which can be truly said, as far we know, to enter into a state of periodical torpidity. Fishes inhabiting ponds and lakes may be occasionally frozen up in ice, when they become stiff, and apparently dead, but upon being cautiously thawed they revive. May we not call these instances, examples of a torpid state induced by accident? However this may be, eels certainly afford an example of periodical hybernation, at least in our colder latitudes. "During the cold months of the year, eels remain imbedded in mud; and large quantities are frequently taken by eel-spears in the soft beds of harbours, and banks of rivers from which the tide recedes, and leaves the surface exposed for several hours every day. The eels bury themselves twelve or sixteen inches deep, near the edge of the navigable channel, and generally near some of the many land-drains, the water of which continues to run in its course over the mud into the channel during the whole time the tide is out. In Somersetshire, the people know how to find the holes in the banks of rivers in which eels are laid up by the hoar-frost not lying over them, as it does elsewhere, and dig them out in heaps. The practice of searching for eels in mud in cold weather is not confined to this country. Dr. Mitchell, in his paper on the fishes of New York, published in the Transactions of the Literary and Philosophical Society of that city, says, 'In winter, eels lie concealed in the mud, and are taken in great numbers by spears.' Thus imbedded in mud in a state of torpidity the eel indicates a low degree of respiration. Dr. Marshall Hall has shown that the quantity of respiration is inversely as the degree of irritability. With a high degree of irritability, and a low respiration, co-exist—First, the power of sustaining the privation of air and food,—Secondly, a low animal temperature,—Thirdly, little activity,—Fourthly, great tenacity of life. All these peculiarities eels are well known to possess. The high degree of irritability of the muscular fibre explains the restless motions of eels during thunder-storms, and helps to account for the enormous captures made in some rivers by the use of gratings, boxes, and eel-pots, or baskets, which imprison all that enter. The power of enduring the effects of a low temperature is shown by the fact, that eels exposed on the ground till frozen, then buried in snow, and at the end of four days put into water, and so thawed slowly, discovered gradually signs of life, and soon perfectly recovered."*

Do eels that continue in salt water, as, for instance, in the Nore, and around the Margate coast, hybernate, as do those which have ascended the river, and attained to the fresh water? We suspect not, but cannot be positive. It is a remarkable fact, that, tenacious of life as the eel is, and capable as it is of being revived from a frozen condition, it is nevertheless highly susceptible of cold. It some-

* Yarrell's British Fishes.

times happens that during intense frosts, accompanied by a piercing east wind, thousands of eels, though buried in the mud, have been destroyed; crawling from their holes in the agonies of death, they have been washed down the stream to the tideway, and thrown upon the beach. Many instances of this kind are on record. In the Arctic regions there are no eels; none inhabit the rivers of Siberia, the Wolga, the Danube, or any of its tributary streams. Few or no eels exist in the mountain streams of the northern portion of our island.

From the vertebrate animals, let us turn to the invertebrate. All our terrestrial mollusca, or snails, and slugs, hybernate. They retire to concealed nooks and crevices, or into the earth, under flagstones, into deep cellars or vaults, and there become torpid. The shelled snails, having fixed upon a convenient retreat, gradually retire within their shell, spreading over the orifice repeated layers of mucus, which harden into a firm tissue, bearing no unapt resemblance to a drum-head in miniature. The common garden snail, (*Helix aspersa,*) before closing the mouth of its shell with a membranous operculum (lid,) glues it by means of mucus to the surface of some object, as a portion of wooden paling, the inside of a garden-pot, the under surface of a bench, or garden seat, etc.; thus secured, it continues till spring, when the operculum is forced away, the shell disengaged, and the animal's self-liberation effected.

M. Gaspard has detailed some curious circumstances attending the hybernation of the edible snail, (*Helix pomatia,*) common on the warmer parts of the continent, and naturalized in Surrey, near Dorking lime-pits. As soon as the first autumnal chills are felt, this species becomes indolent and ceases to eat; it then collects in considerable numbers on banks, in thickets, and in sheltered nooks. Here the groups hide themselves under grass, dead leaves, and the like, and each forms with the anterior part of its muscular foot a cavity, large enough to contain the shell. This being effected, it draws over itself a covering of earth and leaves, and the foot is withdrawn, the collar of the mantle is protruded, and a quantity of air inspired—the orifice of respiration is then closed. When this is done, a fine transparent membrane is formed with its mucus, this being interposed between the mantle and any extraneous substance lying above. The mantle then secretes a quantity of very white fluid over its whole surface, which sets uniformly like plaster of Paris, instantly forming a continuous covering about half a line thick. When this is hardened, the animal separates its mantle from it by another and stronger mucous secretion; and after a few hours, expelling a portion of the air it had previously inspired, it is enabled to shrink a little further into the shell. It now forms additional layers of mucus, expires more air, and thus retires further into the shell. In this way sometimes a fourth, fifth, and even a sixth partition is formed, with intermediate cells filled with air.

The labour of forming the nicely adapted excavation or hybernaculum occupies two or three days, but the whole of October is taken up by the process of closing the shell. About the beginning of April, revivification takes place. The mode of escape from confinement is simple. The air which is contained in the different opercular cells is again inspired, each membranous partition being broken in turn by the pressure of the muscular foot projected through the mantle. When it arrives at the outer calcareous expansion, the animal making a last effort bursts and detaches its most obtuse angle. Then insinuating by little and little the edge of the foot between the shell and the operculum, it forces the latter off, or breaks it away.

How far our fresh water mollusks enter into a state of hybernation is not very palpable; certainly most, if not all, bury themselves during the winter in the oozy mud of rivers, gently-flowing ditches, or drainage courses, lakes, or ponds, and probably become torpid. As it regards marine mollusks, we know little or nothing. It may here be remarked, that in dry weather, even during summer, our terrestrial shelled snails throw a thin operculum over the mouth of their shell, and attach themselves to any convenient substance, and thus appear to become dormant. From this state they are roused by the first warm shower, and creep abroad in the enjoyment of their animal powers. Warm, humid, showery weather, and the approach of eventide, appear to the snail and the slug those combined conditions of atmosphere and light which are most grateful. All are nocturnal.

Spiders, scorpions, and centipedes, etc., hybernate; as also do woodlice, millepedes, etc.; and the same observation applies to numerous insects, not only in their larva and chrysalis state, but in their perfect or *imago* condition. The housefly, the bee, the humble bee, the wasp, the molecricket, the ant, are familiar examples—to which we may add many beetles. Different species pass the winter in different retreats; some crowd together, and like frogs become torpid in associated numbers; some seek a solitary cell. The aquatic beetles burrow into the mud of their pools. Whatever may be their respective winter-quarters, the selected dormitory is admirably adapted to the constitution, mode of life, and wants of the occupant. The degree of cold which insects in their different states while torpid are able to endure is various. The pupæ of the cabbage butterfly, (*Papilio brassicæ*,) exposed to fourteen degrees below the freezing point of Fahrenheit, became converted in a lump of ice; yet these pupæ, after being cautiously thawed, underwent their usual metamorphosis, and became winged butterflies. We must not, however, suppose that the pupæ or chrysalides of all butterflies can sustain such a degree of cold with impunity; yet duly sheltered, the vital principle enables many to endure, not only the winter of our climate, but of more northern latitudes. This shelter the caterpillar ever seeks before changing into a pupa condition, a state which may be fairly regarded as one of hybernation, as well as of preparation for a development of organs hitherto in their germ, a new aspect, and a new mode of life. On the approach of winter, worms plunge deep into the ground, so as to avoid the frost; but the depth to which they burrow depends upon that to which the ground is frozen. We have seen their borings below the alluvial soil, carried some depth into the under stratum of superficial brick clay. Leeches hybernate in the mud. With respect to marine worms, as lug-worms, etc., we cannot speak positively, but we suspect that during winter they remain quiet in their deep borings made in the sand or mud of the beach. So much, then, for a general view of the most remarkable of the phenomena displayed by animal beings.

But if we go to the vegetable kingdom, shall we not there also find a species of hybernation? Annuals perish, but is not the bulbous root of a hyacinth, or tulip, or lily, the hybernaculum of a germ in which the vital principle lies concentrated, protected by envelopes forming a sort of nest? Therein lies the future plant, ready to peep forth, when the severity of midwinter is passed, or even in winter under artificial shelter. Let us look at the trees of our woods and fields—the oak, the sycamore, the horse-chestnut, the elm, the poplar, etc. Is not every sleeping bud itself, as we have said, a distinct organism, enveloped in a scaly covering, sometimes woolly, sometimes smeared with aromatic gum-resin, in a state of hybernation, torpid, until roused into activity by the advance of spring? If the bud begin to open, and a sudden frost set in, it returns not to a state of hybernation, but perishes. "Buds," says sir J. E. Smith, "resist cold only till they begin to grow; hence, according to the nature and earliness of their buds, plants differ in their powers of bearing a severe or variable climate." Looking, again, at the tree as a whole, we find its organic functions suspended. The circulation of the sap, whether in the tubes around the pith, or in the vessels of the bark, or the alburnum, has ceased, or is very languid, but as the spring approaches, the temperature acts on the vital principle, and the sap ascends from the root; the buds now begin to swell, and soon the leaves burst from their scaly investment. The same observations apply to the secretions of plants or trees, whatever the nature of those secretions may be, inasmuch as they are eliminated from the sap, which in its passage through the leaves and bark becomes quite a new fluid, and instead of appearing to the sight and taste like water, as it does when it first commences to rise in spring, possesses the peculiar flavour and qualities of the plant, yields woody matter for the increase of the vegetable body, and furnishes various secretions, as gum, resin, sugar, oil, acids, bitters, essences, iodine, prussic acid, poisonous principles, and what is strange, even silex, or flint.

Thus, then, periodical hybernation, a state of

functional repose, is not peculiar to animals only; we observe it, also, in the vegetable kingdom; the end to be accomplished by the operation of this curious law is the preservation of life, and a preparation of the system for the performance of every function with renewed energy.

CHAPTER VII.

ÆSTIVATION, OR SUMMER SLEEP.

FROM hybernation we now turn to what may, with great propriety, be termed Æstivation. In intertropical climates, a continuance of heat and extreme dryness produce the same effects on animals and plants as does the cold of winter in our temperate latitudes. Mr. Darwin, from whom we borrow the term, in his interesting work, (Voyages of the "Adventure" and "Beagle,") gives us a graphic picture of the stillness which reigns over plain and forest during the hot season of drought. The crocodiles and snakes are all torpid and buried in the ground; the quadrupeds all inanimate; some, as the Tanrec of the Mauritius, become lethargic; the birds seek the dense shade. Not an insect is to be seen. Life appears to stagnate. But no sooner has the rain set in, than the scene becomes changed as if by magic. Those that were sleeping, awake and rouse up, and the aspect of nature is renovated; and not unfrequently has the native tenant of a newly-erected hut or cabin been startled by the shaking of the floor, and the bursting forth of an alligator or serpent. "Sometimes, so the aborigines relate," says Humboldt, "on the margin of the swamps, the moistened clay is seen to blister, and rise slowly in a kind of mound; then, with a violent noise, like the outbreak of a small mud volcano, the heaped-up earth is cast high into the air. The beholder acquainted with the meaning of this spectacle flies, for he knows there will issue forth a gigantic watersnake, or a scaly crocodile, awakened from a torpid state by the first fall of rain." Now all is activity; the air teems with insects; the birds are all busy; the forest resounds with their mingled voices; and the plain is covered with vegetation.

But in order to convey a picture of the whole with due force, we must borrow the pencil of one—a philosopher, and poet-painter—who has himself witnessed the transition, and experienced alike the drought and the shower:—"When under the vertical rays of the never-clouded sun, the carbonized turfy-covering (of the llanos, or steppes) falls into dust, the indurated soil cracks asunder as if from the shock of an earthquake. At such times, two opposing currents, whose conflict produces a rotatory motion, come in contact with the soil, and the plain assumes a strange and singular aspect. Like conical-shaped clouds, the points of which descend to the earth, the sand rises through the rarefied air, in the electrically-charged centre of the whirling current, resembling the loud waterspout dreaded by the experienced mariner. The lowering sky sheds a dim, almost straw-coloured light on the desolate plain. The horizon draws suddenly nearer; the steppe seems to contract, and with it the heart of the wanderer. The hot dusty particles which fill the air increase its suffocating heat, and the east wind blowing over the long heated soil brings with it no refreshment, but rather a still more burning glow. The pools which the yellow fading branches of the fan-palm had protected from evaporation now gradually disappear. As in the icy north, the animals become torpid with cold, so here, under the influence of the parching drought, the crocodile and the boa become motionless and fall asleep, deeply buried in the dry mud. Everywhere the death-threatening drought prevails; and yet, by the play of the refracted rays of light producing the phenomenon of the mirage, the thirsty traveller is pursued by the illusive image of a cool, rippling, watery mirror. The distant palm bush, apparently raised by the influence of its contact with unequally heated, and therefore unequally dense strata of air, hovers above the ground, from which it is separated by a narrow, intervening margin. Half-concealed by the dark clouds of dust, restless with the pain of thirst and hunger, the horses and cattle roam around; the cattle lowing dismally, and the horses stretching out their long necks and snuffing the wind, if haply a moister current may betray the neighbourhood of a not wholly dried-up pool. More sagacious and cunning, the mule seeks a different mode of alleviating his thirst. The ribbed and spherical melon-cactus conceals under its prickly envelope a watery pith. The mule first strikes the prickles aside with his fore-feet, and then ventures warily to approach his lips to the plant and drink the cool juice. But resort to this vegetable fountain is not always without danger; and one sees many animals that have been lamed by the prickles of the cactus.

"When the burning heat of the day is followed by the coolness of the night, which in these latitudes is always of the same length, even then the horses and cattle cannot enjoy repose. Enormous bats suck their blood like vampyres during their sleep, or attach themselves to their backs, causing festering wounds, in which mosquitoes, hippobosces, and a host of stinging insects intrench themselves. Thus, then, these animals lead a painful life, during the season, when under the fierce glow of the sun the soil is deprived of its moisture.

"At length, after the long drought the welcome season of the rains arrives; and then how suddenly is the scene changed! The deep blue of the hitherto perpetually cloudless sky becomes lighter; at night the dark space in the constellation of the Southern Cross is hardly distinguishable; the soft phosphorescent light of the Magellanic clouds fades away; and even the stars in Aquila and Ophiucus in the zenith line shine with a trembling and less planetary light. A single cloud appears in the south, like a distant mountain rising perpendicularly from the horizon. Gradually the increasing vapours spread like mist over the sky, and now the distant thunder ushers in the life-restoring rain. Hardly has the surface of the earth received the refreshing moisture, before the previously barren steppe begins to exhale sweet odours, and to clothe itself with *Kyllingias*, the many panicles of the *Paspalum*, and a variety of grasses. The herbaceous mimosas, with renewed sensibility to the influence of light, unfold their drooping, *slumbering* leaves, to greet the rising sun; and the early song of birds, and the opening blossoms of the water-plants, join to salute the morning. The horses and cattle now graze in the full enjoyment of life. The tall springing grass hides the beautifully spotted jaguar, who, lurking in safe concealment, and measuring carefully the distance of a single bound, springs, cat-like as the Asiatic tiger, on his passing prey." This is, indeed, a forcible picture. How oppressive to life are heat and drought combined! At this season, many tropical plants shed their leaves, and the reptiles bury themselves in the mud. Many mammalia also seek a shelter, and sleep in tranquil torpidity; the land snails are all shut up in their chambers, and take in their snug recesses of retirement their quiet siesta. No insect wing winnows the suffocating air. The worms of the earth plunge deep into the sub-soil, and man himself feels listless, languid, and oppressed. Is this hybernation? In one sense, *yes*—in one sense, *no*. Touch red-hot iron, and it blisters the skin; touch iron in the Arctic regions, while "winter holds her sway," and the same effect is produced. Extremes produce like results. It strikes us very forcibly that the terms *hybernation* and *æstivation* are altogether wrong; they refer to seasons, and not to those peculiarities of animal and vegetable constitution in which sleep for the preservation of life, and the accumulation of vital energy during that sleep, are the results aimed at, so to speak, by the laws of nature. The hot dry season is the winter of the intertropics, if we can use the term winter where no winter is, according to our ideas; and where, in place of rain, of sleet, and snow, and cold, a burning wind, aridity of soil, a drying-up of lakes, rivers, and pools, reduces animal and vegetable energy to its lowest ebb. In the Mauritius, the dry season, when the hedgehog-like tenrac hybernates or æstivates, is 6° 75′ Fahr. above the temperature of the hottest month in Paris. What the heat of Paris is at Midsummer we well know, and so no doubt do many of our readers. It is not, then, cold that produces within the intertropics the torpidity of animal life, but heat and drought; for the rain cools the air, and rouses the torpid slumberers from their lair.

Baron Humboldt expressly alludes to a point which the writer has previously touched upon, in more works than one; and it gives him pleasure to find that his ideas coincide with those of so great a philosopher. Alluding to the tenrac, he says, "As in the cold zone the deprivation of heat causes some animals to fall into winter sleep, so the hot tropical countries afford an analogous phenomenon, which *has not been sufficiently attended to*, and to which I have applied the name of *summer sleep*," (our word *æstivation* means the same.) He then observes, "*Drought and continuous high temperature act like the cold of winter* in diminishing sensibility."

We have already said that we dislike the terms *hybernation* and *æstivation*. They refer to atmospheric temperature; whereas we ought to seek for some term which applies to the condition of the animal itself, requiring (as all animals do more or less, for all hybernate or æstivate, whether they sleep or not) a repose for the invigoration of the mysterious vital principle. Such a term as *life-sleep* (*Hypnobion*—*ὕπνος*, sleep, and *βίος*, life) would not be inappropriate.

CHAPTER VIII.

MIGRATION.

CLOSELY connected with that mysterious vital phenomenon which we call hybernation, æstivation, or, as we have presumed to term it, life-sleep, is that passage of animals from latitude to latitude, from clime to clime, which we commonly term MIGRATION. Migration, however, presents us with several phases, which in general are greatly confounded with each other. Let us arrange them into something like order, so that we may not be troubled or confused in a survey of the differential courses, more or less decided, of organic beings, all impelled by the promptings of an impulsive instinct.

Migration is a movement, under the guidance of instinct, for the accomplishment of certain ends and purposes, of which the actors themselves, however benefited they may be, have not the slightest idea from reflective reasoning—they act as impulse urges. Migration is either perfect and regular, as in the case of the cuckoo, swift, and swallow—or imperfect, as in the case of the kingfisher, which merely leaves the stream for wider water-courses—or irregular, as in the case of the migration of herrings in the north, or of clouds of locusts in Asia or Africa. We will first attend to perfect and regular migration, as we see it performed by "*birds of passage.*"

Our island is well situated for observing the phenomena of migration, which, viewed from it, may be likened to a tide-stream, flowing northwards in spring, with a southern reflux in autumn. We may say, indeed, that our island is an Africa to the wild fowl of the Arctic regions, and an Arctic breeding-place to the swallow, which winters in Africa. Let us paint a summer in the Arctic regions. It is very short, but short as it is, it sees the birth of thousands of most interesting beings, and every islet and every promontory is thronged by a dense population. As if by magic, the snows of winter have dissolved, and coarse herbage has covered the land. Every small pool, every lake, every inlet, is garlanded with vegetation. Driving onwards from the south, (our temperate latitudes,) arrive myriads of wild-fowl, water birds of various species, scoter ducks, widgeons, eider ducks, king ducks, pochards, etc., and also several species of wading birds. The work of incubation now commences. The ground is converted into a city of nests, rarely intruded upon by the foot of man. Here myriads of wild fowl are reared. The water supplies them with food—the reeds bend over their nests—

> ——"by plashy brink
> Of weedy lake, or marge of river wide,
> Or where the rocking billows rise and sink,
> On the chafed ocean-side."

But the summer is, as we have said, short. It passes not into winter by the transition of a mellowed autumn. As it sprang almost of a sudden out of winter, so it retires; but the wild birds, instinct-taught, anticipate the time when river and lake, pond and inlet, will be locked up with ice. Their young are fledged, strong on the wing, and now they commence their southern journey, not to seek a breeding home, but open lakes, open creeks, and seas wherein the ice-floe is never witnessed, and from which they may derive their sustenance. Such are our winter visitors, taking a broad survey of the ornithology of our islands and the adjacent continent.

It is winter in England. In our fens and meres, in our estuaries and lakes, and along our coasts, thousands of wild fowl, harassed, it may be, by the gunner, procure the means of life. But if we have gained, so also have we lost; for on the approach of our winter, the nightingale, and swallow, and wheatear, the quail and the corncrake, and many more, which we will not enumerate, have fled southwards to their winter-quarters. The term, *winter-quarters*, is relative; they seek the congenial atmosphere of a land over which the icy gales of our northern winter are never wafted; but our island is a bounteous store-house—an Africa to thousands of Arctic-reared birds, that are constrained to wing their way southwards when the sea becomes ice-locked, and the land buried beneath a mass of hardened snow. Not only do water birds, or, as they are usually termed, "wild fowl," then visit us, but other species also, which, although permanent residents in our island, are not permanent in more northern latitudes. Among these we may enumerate the thrush, the skylark, the golden-crested wren, the snipe, and the woodcock; other species, again, as the redwing, the fieldfare, the redpole, the siskin, etc., which are truly indigenous in high northern latitudes, conjoin with the former to enjoy the luxury of our comparatively mild winter. It is a remarkable fact that many species of birds, which are permanent residents in our island, are migratory birds in Russia, Siberia, etc., and in this fact we have a clue to the migratory habits of those species which visit our temperate latitudes in spring and depart in autumn.

Let us paint a summer in our island, including of course the temperate latitudes of the adjacent continent. The warm breath of spring has called forth the buds of the forest trees; the snowdrop has given us earnest that winter is about to depart, and the crocus enamels our garden. But not yet does the swallow appear. On the other hand, at this time our winter sojourners are on the move northwards, as if to leave a fair field for their summer successors. The grass of the meadows is springing; the young lambs are sporting by the side of their anxious mothers; a bird dashes past—it is the swallow. The swallow is not, however, our first spring visitor. The wheatear, the redstart, and some others, precede the swallow; but as they come not in vast flocks, their arrival attracts but little notice. But wherefore do they come? Surely in the southern climes, where they take up their winter abode, summer would afford them all that they might need. We may think so, but the God of nature has ordered it otherwise. To our latitudes instinct impels a concourse of summer visitors, as if to make up for the loss of those who have left us for the Arctic regions—" Yea, the stork in the heaven knoweth her appointed times; and the turtle and the crane and the swallow observe the time of their coming," Jer. viii. 7.

It is not here our purpose to enter into a long detail respecting those species of birds which visit us in winter, and depart northwards in early spring, ere the snowdrop rears her head; nor yet shall we cite every species which pays us a summer visit, and departs wending its way southwards in autumn. It is to the great principle by which, unknown even to themselves, they are guided—to an instinctive and a mysterious impulse, that we would call attention. In the consideration of this subject, the law of God is manifest; for we cannot suppose that the bird reasons as man would do were he about to journey to other lands, whether for a temporary visit or for a lifetime. It cannot be doubted that birds of regular migration are, unknown to themselves, directed to avoid two evils—namely, cold and the want of food. So strong, so dominant is this impulse, that migratory birds kept in aviaries, defended from the cold, and plentifully supplied with food, are decidedly agitated when the season of their southern migration comes on. What does the bird require? As far as warmth and food go, nothing. But that impulse which the Creator implanted in it disturbs its frame; it strives to escape from the bars of its prison-house, and failing this, it dies. Hence the great difficulty of keeping the nightingale, the black cap, the wheatear, and others, in confinement. Reptiles awake from torpidity when the temperature is even lower than that of the time in which they retired to their winter-quarters. Our migratory spring birds visit our shores when the cold is lower than it is in autumn, when they take their departure. At the period of their return, however, the vital principle has acquired as it were accumulated force by repose; their animalization seems to be at its highest point; and in like manner, their insect prey, roused from torpidity, or emerging from the pupa state, is ready for their acceptance. Throughout the whole chain of nature, vitality is refreshed by repose, and this repose is experienced in some animals by hybernation, in others by change of residence.

As some birds, which in the north are migratory, may be regarded as stationary in our latitudes, so some species, which are migratory in our island, remain stationary in southern

Europe. In England, the thrush, the skylark, etc., are stationary; but every winter, especially if the season be severe, the flocks of our home-bred birds are increased by arrivals from the north. In northern Germany, Denmark, Sweden, and Norway, the redbreast is migratory, departing southwards in autumn. In our island, in France, and throughout southern Europe generally, it is stationary; as it is also in south-western Asia, and northern Africa. In Portugal, the quail appears to be stationary; but the quails which visit our island and the adjacent parts of the continent, move southwards in autumn, visit the islands of the Mediterranean, and pass to the shores of northern Africa. In the northern parts of our island, the swallow, the cuckoo, the goatsucker, and others, are from a week to ten days later before making their appearance than they are in our southern counties; and what is singular, some birds never visit either our western or northern counties, or at least very rarely. The nightingale, for example, which annually visits northern Germany, and even Sweden, is rare in Yorkshire, and unknown further north; its appearance in Wales, in Devonshire, in Cornwall, is of rare occurrence, while in Ireland it has never been seen, as far as we can learn. Its rival in melody, the blackcap, however, is a summer guest in Scotland. In Madeira, the latter bird is a permanent resident. In like manner, those birds which leave our island in autumn, reside longer in the southern counties than they do in the northern. Thus, dutterels forsake the Grampians about the beginning of August, and quit Scotland at the end of that month, while they remain about our southern shores to the latter part of September, or even the beginning of November. Dr. Fleming states, that a difference of nearly a month takes place between the departure of the goatsucker from Scotland and the south of England.

It must not be supposed that migratory birds perform their voyage by one continuous flight to the place of their winter rest, or of their summer sojourn. They rest by the way, and many, as the swallow and the wheatear, travel during the night. At the same time, they cannot spend much time in rest, for Adanson observed the swallow in Africa in the month of October, after its return from Europe—most probably from the countries bordering the Mediterranean. M. Prelong says, "I have observed, as Adanson also has, that our swallows and our wagtails arrive in the torrid zone eight or ten days after the period of their departure from our climates. In 1788, I saw the yellow wagtails, (*Bergeronettes du printems,*) and the grey wagtails, on their arrival at Goree on the 14th of September. Adanson says that he has seen swallows arrive at Senegal on the 9th of October; now I recollect that they quit the department of the High Alps about the end of September—thus making the time to tally."*

The migratory movement is far more exten-

* *Annales de Chimie,* tom. xvii. p. 272.

sive in some cases than in others, that is, there is a greater distance between the summer and winter abodes of certain species than of other species, which at the same time migrate regularly. For example, the gannet, which breeds in flocks on the Bass Rock, on St. Kilda, on the Isle of Ailsa, on the Skelig Isles, etc., spends the winter in the Bay of Biscay and in the Mediterranean, feeding upon the anchovy and sardine. The shearwater, (*Puffinus Anglorum*,) which breeds abundantly in the Orkneys, arriving in February or March, visits in winter the coasts of Spain and the Mediterranean. The guillemot and razor-bill, which breed on the Needles (Isle of Wight) in great numbers, also resort to the same winter-quarters.

Our island, then, favourably situated, receives an autumnal influx of birds from the north—wild fowl, woodcocks, snow-buntings, redpoles, etc., which pass the cold season with us, and which in spring retire northwards, giving place to our summer guests. The latter, inasmuch as they breed with us and not in their winter asylum, may be regarded as truly indigenous; while the former, which make our island a convenient retreat, but breed in the Arctic Regions, cannot be considered as truly indigenous, though they occupy a place in the list of British birds. Of these, however, some few do breed in the northern districts of our island, and are thus borderers:—the redpole, (*Linaria minor*,) and the siskin, (*Carduelis spinus*,) are examples in point.

Much has been written respecting the migration of birds, treating the subject as if the birds themselves reasoned on the necessity of their journey and on the course most advantageous for them to take, as the captain of a vessel bound on the one hand for the high-northern whale-fishery, or on the contrary for Australia or the southern seas, might do, with his chart before him, calculating the effects of currents and trade winds. Herein many persons labour under a singular confusion of ideas ; the end to be answered by the migration of a bird is undoubtedly its preservation, but the mode of operation is instinctive. The bird knows nothing of north, south, east, or west ; yet as instinct directs it, so it urges its flight, and but for this instinctive impulse it would remain stationary, and in many instances starve.

So far we have attended only to the instinct-directed migration of birds. But we must not forget that other animals regularly migrate, and here we more particularly allude to the oceanic mammalia. The movements of these, however, as may be easily imagined, are only partially known to us. For example, certain seals in the Arctic regions certainly migrate, while others are stationary, or only perform a short journey, as circumstances connected with their fishing operations may render convenient. Crantz, in his History of Greenland, notices two species of seal which he considers as decidedly migratory. One he calls Neitersoak, (its Esquimaux name,) which we take from his description to be the *Phoca Cristata;* Desmarest, (*P. leonina;* Fabr.;) and the other, Utsuk, apparently the *Phoca barbata* of Müller. These species emigrate twice in the year from the coasts of Greenland. First, they retire in July and return again in September, and this expedition southwards is apparently undertaken, as Crantz supposes, in quest of food, but other motives are probably influential. The time of their second departure is in March. They return in the beginning of June, the females with their young, crowding together like flocks of sheep. After their first voyage they come back in good condition, but when they appear in June they are very lean.

"In this last tour," says Crantz, "they seem to observe a certain fixed time and track, like the birds of passage, and take a route that is free from ice; therefore the ships near Spitzburg can safely follow them. We know they come up out of the south first, then twenty days after they are eighty or one hundred leagues further north, and the longer the date the further they lose themselves in the north. We can pretty well ascertain the day at the end of May when they will be again at Frederick's Hope, and in the beginning of June at Good Hope, and so further north. But the place they retire to in that last circuit cannot be determined with equal certainty." . . . "They do not go to America, for their course is not steered westward, but northward; nor do the sailors ever see them in the open sea at this season. They do not stay in the north; they do not breed there, either on the ice or among the secluded rocks, for we see them return from the south, and not from the north; therefore they must either find a way through some narrow passage or sound, such as a certain channel may prove to be in Disko Bay, (69° north latitude,) or that in Thomas Smith's Sound, (78°,) or else they must get round Greenland, through some supposed open sea further north under the pole, and so arrive at the east side, and then pass between Iceland and eastern Greenland and round by Statenhook." Crantz wrote from observation, without pretending to be a professor of zoology; and to simple narratives, written without bias or predilection for any favourite theory we are constrained to pay just deference. He says that the harp seal, or attarsoak, is only, if we understand him rightly, partially migratory from Greenland. Desmarest, however, speaking of this species, (*Phoca Greenlandica,*) states, that it leaves the coast ot Greenland twice a year, in the month of March to return in May, and again in July, to return in September. Professor Bell, in his History of British Quadrupeds, makes the same observation, adding that these seals "remain principally on the floating masses of ice, and are thus occasionally carried to various parts of the northern sea. In the same manner, it is probable that some unusual drifting of such floes of ice might have been the means of transporting the individuals which have sometimes been found in the Severn, into a more southern latitude than ordinary, from whence they made their way to that neighbourhood." The appearance of the harp seal in our seas is merely accidental ; it has been shot near the Orkney Islands. The probability is, that all seals are more or less migratory. The elephant seal (*Phoca proboscidia*) of the Southern Ocean, according to the best information which we possess, travels between latitude 35° and 50°, passing to the Antarctic in summer, and retracing its course in winter.

It has been asserted by many that fishes perform extensive migrations ; but perhaps it will be found that their migratory journeys, though regular, are far more limited than has been supposed. For example, the mackerel was believed by Anderson, Duhamel, and others, to be a fish of passage, performing, like some birds, certain periodical migrations, and making long voyages from north to south at one season of the year, and the reverse at another. Mr. Yarrell combats this error, observing—"It does not appear to have been sufficiently considered, that inhabiting a medium which varies but little either in its temperature or production, locally, fishes are removed beyond the influence of the two principal causes which make temporary change of situation necessary. Independently of the difficulty of tracing the course pursued through so vast an expanse of water, the order of the appearance of the fish at different places on the temperate and southern parts of Europe is the reverse of that which, according to this theory, ought to have happened. It is known that this fish is now taken even on some parts of our own coast, in every month of the year. It is probable that the mackerel inhabits almost the whole of the European seas; and the law of nature, which obliges them and many others to visit the shallower water of the shores at a particular season, appears to be one of those wise and bountiful provisions of the Creator, by which, not only is the species perpetuated with the greatest certainty, but a large portion of the parent animals are brought within the reach of man, who, but for the action of this law, would be deprived of those species most valuable to him as food." The salmon is, to a certain extent, migratory. It leaves the adjacent sea for the estuary, and the estuary for the river—the motive for this change being the selection of a fresh-water breeding-place; for although the salmon is a marine as well as river fish, it commences existence in the clearest portion of inland streams; and what is more, it never migrates far, in its ocean-ward visit, from the embouchure of its native river. The period of the year at which the salmon enters the estuary and ascends the river, differs according to the temperature of a given river, which temperature is influenced by many circumstances. Some rivers are filled with salmon far earlier than are others. In the autumn, or early part of winter, salmon recede from the sea, and enter the estuary of their chosen river; here they lie, advancing with the flood tide, and retiring with the ebb. As the season progresses, they ascend higher up the river, so as to escape the influence of the tide, and are now getting into full roe. "Their progress forwards is not easily stopped; they shoot up rapids with the velocity of arrows, and make wonderful efforts to surmount cascades and other impediments by leaping, frequently clearing an elevation of eight or ten feet, and gaining the water above pursue their course. If they fail in their attempt, and fall back into the stream, it is only to remain a short time quiescent, and thus recruit their strength to enable them to make new efforts." The salmon deposits its spawn in September, October, and November, and returns to the sea. "With the floods of the end of winter or commencement of spring, the salmon, male and female, descend the river from pool to pool, and ultimately gain the sea, where they quickly recover their condition, to ascend in autumn for the same purpose as before." The descent of the young fry to the sea begins in March, and continues through April and May. Here, then, we have an example of periodical migration, but on a limited scale; nevertheless, we cannot deny that it is true migration—a migration from the deep sea to the highly aërated water of the pure river.

As we are now attending to the migration of fishes, which, we contend, is very partial, even where most regular, we will look at that fish, which, in the imagination of our older writers, is a wonderful traveller—we mean the herring. In our younger days, with what interest did we not trace the course of this fish from the Polar regions to our shores! But in science, if our mind be open, truth either steals upon us by degrees, or bursts upon us suddenly, making us smile at our former ignorance. According to Pennant, Anderson, and others, the great winter rendezvous of the herring is within the Arctic circle, where, as the former author (a man of great attainments) observes, they continue for many months, after the exhaustion of the system consequent upon reproduction. He adds, that the sea within that space, swarming with insect-food in a far greater degree than in our summer latitudes, is consequently attractive to the herring tribes. Thus, on the arrival of spring, according to Pennant, "the mighty army begins to put itself into motion," and commences a southward course. We find, however, in Pennant a little inconsistency; for, acknowledging that herrings continue on the Welsh coast till February, that is throughout the winter months, he tells us that the great horde, turning back from winter-quarters, begins to appear off the Shetland Islands in April and May. "This," he continues, "is the first check (land check) which the army meets with on its march southward. Here it is divided into two parts; one wing of those destined to visit our coast takes to the east, the other to the western shores of Great Britain, and fill every creek with their numbers. Others proceed towards Yarmouth, the great and ancient mart of herrings; they then pass through the British Channel, and after that in a manner disappear. Those which take to the west, after offering themselves to the Hebrides, where the great stationary fishery is, proceed towards the north of Ireland, where they meet with a second interruption, and are obliged to make a second division. One branch takes to the western side, and is scarcely perceived, being soon lost in the immensity of the Atlantic; but the other, which passes into the Irish Sea, rejoices and feeds the inhabitants of most of the coasts that border on it. These brigades, as we may call them, which are thus separated from the greater columns, are often capricious in their motions, and do not show an invariable attachment to their haunts."

It would seem as if Pennant had been the leader of a wing or brigade of the herring army, so precisely does he describe the course of advance from north to south. But alas, like many theories entertained by men of high intellect and observant minds, Pennant's history of the fish in question, as far as its army movements are concerned, is altogether visionary. We deny not that the herring migrates, but it is from the deep sea to shoal-water, and *vice versâ*—and this is all. The herring is never seen in the Arctic seas. The whale fishers do not meet with it in high latitudes. Crantz observes, that in the south of Greenland some few of the *large sort* of herrings are taken, "wanderers that have strayed from the great shoal that drives out of the ice-sea by Iceland towards America." He regarded, as might be expected, the theory of the Arctic migration of the herring as correct. But he previously observes, that the *proper herring* is not an Arctic fish. There is, however, a distinct species, peculiar to the northern ocean, called by the Greenlanders *Angmarset,* or small herring, and by the Newfoundland-men, capelin, which was observed by sir John Franklin along the shores of the Polar sea. Whatever this species may be, it is not the true herring, or *Clupea harengus.**

The herring, then, and the salmon also, afford us instances of *imperfect migration.* And here we may adduce some other examples, by way of showing, what all do not seem to understand, that migration does not necessarily involve a great change of latitude, (which we have called *perfect or complete migration,*) but only a removal from a given spot, for a definite purpose, to the extent merely of one or two hundred miles. Referring here to our work on British Birds,† we shall select such illustrations as have come under our personal knowledge, although we may corroborate our opinions by the authority of the most eminent zoologists. As the first example of imperfect or limited migration, we

* See Yarrell's British Fishes.
† Published by the Religious Tract Society.

may cite the lapwing, or peewit (*Vanellus cristatus.*) During the summer, this bird tenants heaths, open commons, moorlands, morasses, and turf-bogs, etc., where it rears its young. It is abundant in the moorlands of the Peak of Derbyshire, appearing early in spring. In the autumn it collects into flocks, which depart sooner or later, according to the severity of the weather, not having, as far as we ourselves have observed, any decidedly fixed time for leaving their summer quarters. These flocks wing their way to the lowlands and marshes bordering the estuaries of our larger rivers, where, from the oozy ground, they obtain a supply of food. We have seen extensive flocks along the estuary of the Thames, both on the Kentish and Essex coasts, and no doubt some continue in the marsh lands along the Thames throughout the whole of the year.

Dr. Fleming tells us that in Scotland the lapwing, although it visits the open fields near the coast in winter, wings its way southwards when the weather sets in with severity, and may be regarded as a stationary bird in England, of course changing its locality from the moorland or heath to the marsh lands along the shore. Here, then, we have an example of limited migration, as performed by a bird diffused over Europe and Asia, the end to be accomplished being, not an escape from cold, but the necessity of obtaining food.

As another example, we may refer to our kingfisher, noted alike for the beauty of its plumage and its fish-feeding habits. In summer, as is well known, this bird frequents our inland rivers, breeding in holes on the bank; but "when winter sets in, and drives the fish from shallows to deep and sheltered bottoms, freezes the mill-dams, or coats with ice the sluggish basin, worked out by the current in rich alluvial soil, these birds wander from the interior to the coast, and frequent the mouths of rivulets entering large navigable rivers, dykes near the sea, and similar places, especially on the southern portion of our island."

Passing from imperfect or limited migration, conducted with regularity, we next advert to an irregular migration, performed by various animals as circumstances may influence them, and of which among mammalia the lemming affords us an example. This little rodent, a sort of field rat, is a native of northern Russia, Norway, Sweden, and Lapland, etc., and has been long notorious for its incursions upon the cultivated districts bordering its native mountain home. The ordinary food of the lemming consists of grass, the reindeer lichen, the catkins of the dwarf birch, etc., to which they add animal food when obtainable, sometimes even destroying and devouring the weaker of their own species, impelled, as we believe, by extreme hunger. It would appear that they accumulate at times in their native fastnesses, and are prompted by instinct to migrate from an over-populated territory. This migration takes place once or twice in the course of fifteen or twenty years.

Great bands descending from the Kiölen mountains traverse Nordlands and Finmark, ending their journey and their lives in the Western Ocean, into which host after host rushes, there to perish. Others, taking a direction through Swedish Lapland, are drowned in the Gulf of Bothnia. Their march is stated to be in parallel lines, about three feet apart, without stop or stay unless the obstacle be insurmountable. They cross rivers and lakes without deviation, and pouring onwards, to the consternation of the cultivators of the soil, spread over the land, devouring every green thing, and are said to gnaw through corn and hay-stacks, leaving desolation in their track. "They march," says Pennant, "like the army of locusts, so emphatically described by the prophet Joel, destroy every root of grass before them, and spread universal desolation. They infect the very ground, and cattle are said to perish after tasting the grass which they have touched. They march by myriads in regular lines; nothing stops their progress; neither fire, torrents, lake, nor morass. They bend their course straight forward with most amazing obstinacy; they swim over lakes; the greatest rock gives them but a slight check; they go round it, and then resume their march directly on without the least diversion. If they meet a peasant, they persist in their course, and jump high as his knees in defence of their progress. They are even so fierce as to lay hold of a stick, and suffer themselves to be swung about before they quit their hold. If struck, they turn about and bite, and will make a noise like a dog." This description, though it may be amplified even to the borders of romance, is doubtless true in its essentials.

We may next allude to the reindeer, as affording us a fair example of regular although limited migration. In the Arctic regions of Europe and Asia, this animal in its wild condition herds in troops, which travel from the woods to the open hills, and back again according to the season. The woods are their winter refuge; here they subsist on the long pendent lichens which hang in festoons from the trees, on the white lichen which covers the ground, and on the twigs of the birch and willow. With the return of spring they begin their migration from the forest to the mountain ranges, partly in order to obtain their favourite food, and partly to escape the heat and the countless myriads of mosquitoes which then begin to swarm, as well as the gadfly, a dreaded tormentor. So imperative is the instinct that impels the reindeer to these migratory movements, that it cannot be modified in the domestic race, which constitutes the chief wealth of the Laplander. Hence the Lapland peasant is obliged to lead a semi-nomadic life, taking periodical journeys of no common toil, from the interior of the country to the mountains which overhang the Lapland and Norway coasts, and thence back to the interior. The reindeer, or caribou, in the high regions of

North America, is migratory, like the European animal. Dr. Richardson describes two varieties, the woodland caribou, and the barren-ground caribou. Both have their peculiar winter and summer localities. The former exceeds the latter in stature; its proper country is a strip of low primitive rocks, well clothed with wood, above a hundred miles wide, and extending at the distance of eighty and a hundred miles from the shore of the Hudson's Bay, from Lake Athapescow to Lake Superior. "Contrary to the practice of the barren-ground caribou, the woodland variety travels to the southward in the spring. They cross the Nelson and Severn rivers in immense herds in the month of May, pass the summer on the low and marshy shores of James' Bay, and return to the northward, and at the same time retire more inland in the month of September."

On the other hand, the herds of the barren-ground caribou spend the summer on the coasts of the Arctic Sea, and in winter retire to the woods between 63° and 66° of latitude, where they feed on the tisneæ, alectariæ, and other lichens, as well as on the long grass of the swamps. About the end of April, they make excursions from the woods in order to obtain the terrestrial lichens, which, now that the snows are partially melted, are both soft and easy to be collected. "In May, the females proceed to the sea-coast, and towards the end of June the males are in full march in the same direction. At this period, the sun has dried up the lichens on the barren grounds, and the caribou frequents the moist pastures, which cover the narrow valleys on the coasts and islands of the Arctic Sea, where they graze on the sprouting *carices*, and on the withered grass or hay of the preceding year, which at that period is still standing and retaining part of its sap. The spring journey is performed partly on the snow, and partly after the snow has disappeared, on the ice covering the rivers and lakes, which have in general a northernly direction."

The North American bison is also migratory in its habits, but its migratory movements appear to be irregular. Herds, often consisting of many thousands, wander over the prairies watered by the Arkansas, Plata, Missouri, and the upper branches of the Saskatchewan and Peace rivers, changing their pasture grounds, and often visiting the salt springs or saline morasses.

Among insects, the swarming and departure of bees from the old hive in quest of a new habitation may be considered as an irregular migratory movement, and so may the swarming and flight of clouds of ants issuing forth to found a new colony.

The locust also affords us a remarkable example of irregular migration as performed by insects. We might here fill some pages with details, but the subject has been so fully discussed, and so many popular accounts have been given, that we shall not trouble our readers with a narrative of them.

As we have been reviewing, in a cursory manner, the laws of migration, selecting by way of illustration the habits of various animals, with which we are more or less familiar, we would here add an observation relative to the periodical, but limited migrations of certain species of crabs, (or *crustacea,*) which are not destitute of considerable interest. Many crabs (as those of the genera *Ocypoda*, *Gelasimus*, etc.,) are semi-aquatic; but some are much less aquatic than others, and some not *only migrate*, but *also hybernate*. As an example in point, we may first notice the sand-crab of Catesby, (*Ocyoda arenaria,*) a native of the coasts of America and the Antilles. The habits of this species are very singular. During the whole of the summer the animal lives on the sea-shore, where it excavates for itself a burrow three or four feet in depth, above the highest line of the tides, or even the dash of the waves. In this burrow it secludes itself during the day, but comes abroad on the approach of dusk, to roam in quest of food. It now visits the shore, traverses the wet sands, and dabbles in the pools. Upon any alarm, it scuds away with great rapidity, at the same time elevating its claws, as if to warn the pursuer of its resolution to defend itself. Towards the end of October, troops of these crabs leave the sea-side and journey inland, in search of some suitable spot where, like marmots, they may excavate deep burrows in which to hybernate. Into these burrows they retire, and stop up the entrance with such address, that no appearance of an excavation is, at least upon a casual survey, to be perceived. This effected, they betake themselves to the bottom, and remain quiescent until the return of spring, when they emerge from their dormitories, retrace their journey to the sea-side, and work out their summer habitations. A species of *gelasimus*, a native of South Carolina, was observed by M. Bosc to migrate inland, and form a burrow, in which retreat three winter months are passed. In fact, according to that naturalist, this species does not seek the water, until the period of depositing its eggs.

But there is a group of crabs still more terrestrial and more migratory than those to which we have alluded. These crabs belong to the genus *Gecarcinus*. They are known under various names, as land-crabs, *toulouroux*, *crâbes peints*, *crâbes violets*, etc. So decidedly are their branchiæ or gills organized for aërial respiration, that if submerged for any length of time in the sea, they will perish from suffocation; nevertheless, it is essential that the branchiæ be always kept moist, for death, on the other hand, results from the desiccation of these organs.* The land-crabs are distributed through the warmer regions of the Old and New World, and are also found in Australasia; but the species are most numerous in America and its islands.

* The common woodlouse, *Oniscus murarius*, or *Cloporte* of the French, within the pale of the crustacea, requires a humid locality for its existence, but is drowned in water.

From various sources we gather that they live more or less inland, paying, at stated periods, a short visit to the sea; the females for the purpose of disencumbering themselves of their eggs, which are carried on the under surface of the body. On land they dwell in burrows, and in these burrows they undergo the process of exuviation—that is, they cast off their old shells, and acquire a new coat of armour. M. Latreille, the accomplished assistant of baron Cuvier, in his *Règne Animal*, sums up the life of these land-crabs in the following words: "They pass the greatest part of their life on land, hiding in burrows, whence they issue forth in the evening. Some take up their abodes in graveyards. Once a year, when they would lay their eggs, they assemble in numerous troops, and take the shortest course to the sea, without being deterred by any obstacles which they may meet with on the road. After the deposition of the eggs, they return in a state of great debility. During the season of exuviation, they block up the mouths of their burrows. Whilst undergoing this process, and while still soft, they are termed *Boursiers*, or purse-crabs, and their flesh is then held in high estimation. Nevertheless it is sometimes found to be deleterious, and this quality is attributed to the fruit of the manchineel, (*Mancenillier*,) of which it is supposed, perhaps without foundation, that they have eaten."

M. Milne Edwards (*Hist. Nat. des Crustacées*),

states in general terms, that for the most part these land-crabs haunt humid places, and conceal themselves in holes which they excavate in the earth; but the localities preferred vary according to the species. Some give preference to low and marshy lands bordering the sea, while others choose hills covered with wood at a considerable distance from the shore. The latter, however, at certain epochs, leave their habitual dwelling, and migrate to the sea. These crabs, as it is reported, unite in great bands, which perform journeys of no trifling extent, laying waste everything in their progress, unstopped by any (but a naturally insurmountable) object. Vegetable substances compose their principal diet, and their habits are nocturnal or *crepuscular*.

It is more particularly in the rainy season that they quit their burrows, in which during their moult they remain closely concealed. They run with great celerity. Browne, speaking of the land-crab (black crab, violet crab, *Gecarcinus Ruricola*) of the Antilles, observes that "these creatures are very numerous in some parts of Jamaica, as well as in the neighbouring islands, and on the coast of the main continent. They are generally of a dark purple colour, but this often varies, and you frequently find them spotted, or entirely of another hue. They live chiefly on dry land, at a considerable distance from the sea, which, however, they visit once a year to wash off their spawn, and afterwards return to the woods and higher lands, where they continue for the remaining part of the season; nor do their young ones ever fail to follow them as soon as ever they are able to crawl. The old crabs generally regain their habitations in the mountains, which are seldom within less than a mile, and not often above three from the shore, by the latter end of June, and then provide themselves with convenient burrows, in which they pass the greater part of the day, going out only at night to feed. In December and January they begin to be in spawn, and are then very fat and delicate; but they continue to grow richer until the month of May, which is the season for them to wash off their eggs. They begin to move down in February, and are very much abroad in March and April—the pairing season." About the beginning or middle of June, the inland migration commences; and "about the months of July or August the crabs fatten again, and prepare for *mouldering*, (that is, retreating to their burrows,) filling up the holes with dry grass, leaves, and abundance of other materials. When the proper time comes, each retires to its hole, shuts up the passage, and remains quite inactive, until it gets rid of its old shell, and is fully provided with a new one. How long they continue in this state is uncertain, but the shell is observed to burst both at the back and sides, to give a passage to the body. Afterwards, the animal gradually extracts its limbs from all the other parts. At this time, the crab is in the richest state, and is covered only with a tender membranous skin, variegated with a multitude of reddish veins; but this in time hardens, and soon becomes a perfect shell like the former. This crab runs very fast, and always endeavours to get into some hole or crevice on the approach of danger; nor does it wholly depend on its art and swiftness, for while it retreats it keeps both claws expanded, ready to catch the offender if he should come within its reach, and if it succeeds on these occasions it commonly throws off the claw, which continues to squeeze with incredible force for nearly a minute afterwards; while it, (the crab,) regardless of the loss, endeavours to make its escape, and gain a more secure or more lonely covert, contented to renew his limb with his coat at the ensuing change."

Turning from the New World to India, we may refer to some details which are given by the late bishop Heber, relative to some land-crabs which he met with at a great distance from the sea, all access to which appeared to be denied by formidable obstacles. Most probably, however, these crabs were fluviatile in their habits, (as is the river crayfish of Europe, the miniature representative of the marine lobster,) and deposited their eggs in adjacent tanks or rivers. If our suspicion be correct, there will be nothing strange in the fact that fluviatile as well as marine crustacea are semi-terrestial. Let us, however, read what bishop Heber details:—" The plain of Poonah is very bare of trees, and though there are some gardens immediately around the city, yet as both these and the city itself lie in a small hollow on the banks of the *river Moola*, they are not sufficiently conspicuous to intercept the general character of nakedness in the picture, any more than the young fir-trees and shrubs with which the bungalows of the encampment are intermingled. The principal and most pleasing feature is a small insulated hill immediately over the town, with a temple of the goddess Parvati on its summit, and a *large tank* (which, when I saw it, was nearly dry) at its base. All the grass land round this tank, and generally through the Deccan, swarms with a small land-crab, which burrows in the ground, and runs with considerable swiftness, even when encumbered with a bundle of food almost as big as itself. This food is grass, or the green stalks of rice, and it is amusing to see them sitting, as it were, upright to eat their hay with their sharp pincers, then waddling off with their sheaf to their holes as quickly as their sidelong pace will carry them."

With these illustrations of our subject we conclude.

Some one, however, may say—Among all the migratory animals to which you have adverted, do you not include man, who has spread his race over the whole of the habitable globe? We answer—No. Migration in the lower animals is the result of an instinctive impulse which they cannot but obey. In the seeds of plants, also, erratic migration is involuntary, although appointed and governed, both as to time and extent, by the predetermined plan of Wisdom in creation. To man it was appointed to replenish the earth and to subdue it; and in fulfilling this decree in all ages, man uses his judgment as to whether he shall quit his native home or not, and also as to the new home in which he shall settle. With him the whole affair is voluntary; he is not instinct-driven. He fits out ships or vessels of some kind, he explores distant regions, having various motives for so doing, and sometimes perhaps is cast away on a distant shore—driven thither by the stress of winds and waves. Where inducements incline his will, he founds colonies; where the prospect of gain invites him, thither he repairs; and not unfrequently has one portion of mankind been forced, by the irruption of fierce hordes of barbarians, exulting in conquest, and intent upon the extension of territory, to retreat before them, leaving their country a spoil. Thus have the waves of the human race rolled on from east to west, from north to south—nations blending with nations—the conquerors with the conquered — languages fused together — new nations and new languages becoming developed in the course of time, till, in the contemplation of the subject, the student in history and philosophy becomes lost in a maze of confusion—"a mighty maze, but not without a plan,"—for every movement of this complex machine, this strife and turmoil of jarring elements, this flux and reflux, this crush of empires, this rise of others on their ruins, this development of might from apparently inadequate beginnings, this change of the great centres of civilization—all are under the control of an Almighty God, who, in the depth of his wisdom, in his secret counsels, has, so to speak, fashioned out every event, so that nothing comes to pass by the working of *blind chance*—a term which even the philosophy of the sceptic must reject.

Passing from migration, there is another point in the ecŏnomy of animals which is worthy of notice—we allude to the habits of such species as are called *social* or *gregarious*, terms which we deem of very different import. Let us then proceed to explain our views.

Animals are either solitary, gregarious, or social. By the term *solitary*, we mean a mode of life alone and apart (excepting with a mate) from others of its species. Under the term *gregarious*, we include such animals as merely associate together, each individual working for itself alone. Under the term *social*, we class together animals which not only congregate together, but which give their joint labours for the common good. Now, although this broad distinction is palpable, looking at given points without transitional degrees, it is not always easy to determine where the line of distinction is to be drawn. For example, numbers of our migratory birds are solitary during the summer, the pairs living apart from other pairs of the same kind; but as the time of autumnal migration comes on, numbers congregate together, and take their south-eastern course. In flocks, likewise, do they return in spring, to separate into pairs on their arrival. Some birds, as the linnet, which during the summer is solitary in our sense of the term, are gregarious during the winter, as, for example, the fieldfare and the starling. On the contrary, the robin is solitary and pugnacious, defending its territory from the encroachment of a rival. But under what head shall we class the polygamous birds, as the ruff, the capercalize, and the wild turkey—species, the males of which are solitary, the females under his governance partly gregarious, while as autumn comes on, males, females, and the young birds of the year associate in families together? or the partridge, which pairs with a single mate, but associates in autumn and winter, not only with its own brood, but often unites that with other broods, so as to form a large covey? In these and similar instances we may call the species *temporarily gregarious*. Such are also the lapwing, the wheat-ear, and many more.

But, perhaps, we ought to observe that under the term *gregarious* two different meanings are to be understood. There is voluntary and involuntary association. The rook, the starling, the antelope, are voluntarily gregarious—their own instinctive nature inclines them to be so; but, on the other hand, the character of the soil, of the water, of the atmosphere, causes a congregation of the same species in certain localities, wherein multiplication goes on because those localities are favourable. Such animals are *congregating*, not *gregarious;* they do not know each other as every rook knows its fellow, every partridge the members of its own covey; they merely multiply and live together, but do not associate—nay, they are often at war with each other, the larger preying on the smaller. It strikes us that in this light we must look at the leeches of the quagmire, the worms of the garden mould or of the fields, at the beds of cockles, mussels, and oysters on our coast, at the dancing maze of gnats floating in the air, and at the fishes which live in shoals in the water. The same observation applies to birds which migrate in vast flocks collected for one common purpose from many districts. Humboldt terms many plants *social*, such as the *Fucus natans*, which covers leagues of ocean —the heaths, which in dense array spread over our hilly moors—the duckweed (*Lemna*) of the sluggish stream—the pine of the northern forest; but in these cases we contend, first, that *social* is not a proper term, and next, that the peculiarity of soil, water, climate, inducing their multiplication, prevents the intrusion of other plants, that very multiplication tending in its turn to repel their intrusion. Plants, therefore, are not *social* nor *gregarious—congregating* they may be, only because they find a congenial spot for development. Beings which are gregarious seek each other out, whether for a season or for a permanent union—beings which are social, not only form assemblages, but labour conjointly, instinct-directed, to effect some definite purpose. Instinct does not bring plants together, but it does make animals seek each other's society. The seeds of a plant fall around it—if the soil and air be congenial, a congregation is the result; but *truly gregarious* beings seek each other out, and form a body politic. Now, shoals of fishes, hordes of crustacea, springing from multitudes of eggs deposited in a given locality, may to a certain extent be compared to what baron Humboldt terms *social* plants. Yet at the same time there is this difference; they instinctively associate together for the purpose of visiting feeding-grounds in rivers, or of migrating from the deep sea to shallower water near the shore. Some fishes, however, may be regarded as solitary, although of these the young fry congregate together in multitudes; but even here the congregation is rather the effect of circumstantial necessity than of choice. A colony of rooks or herons presents us with a a well-ordered republic, but this cannot be said of a shoal of mackerel or herrings, nor even of a cloud of locusts. At the same time, it is difficult to draw a strict line between animals which congregate together and those which are gregarious, for we can scarcely tell how far association is coveted for the sake of association alone, or how far instinct leading animals to the same ultimate objects renders their association rather necessary then voluntary. For example, instinct-directed fishes may herd together in multitudes without any unity among the individuals; but in a rookery this is not the case—every individual of the community is known to all the rest, and a stranger is repulsed; yet, although the rook is gregarious, (and we may say the same of the jackdaw,) their allies, the raven, the crow, the magpie, and the jay, are solitary.

Looking at the animal kingdom as a whole, we should say that *congregational* habits were most prevalent among the lower orders, *gregarious* habits among the higher orders, while *social* habits are allotted only to the few. As examples of the first, we may cite fishes in general, (of which, however, some are solitary;) among insects, locusts; and among birds, the gannet, razor bill, puffin, and many other sea birds. As examples of the second, we may instance the rook, the heron, the wild sheep, the springbok, antelope, and the quagga of South Africa, for here we find a commonwealth of individuals communicating with each other for mutual benefit, each at the same time acting for itself alone. As examples of social animals, we may cite among insects the honey bee, the ant, the termite, the wasp, and others; and among birds, the South African weaver bird, (see "Popular Ornithology," published by the Tract Society;) among quadrupeds, perhaps the seals, the beaver, the musquash, and some few others. As regards solitary animals, we may observe that, as a rule, all predatory or ferocious animals are solitary, (of course, we mean in pairs,) but by way of exception the wild dog hunts in packs, and so does the jackal. The rat, which is partly carnivorous and partly frugivorous, is, to a certain extent, a social animal. At the same time we would only call those animals *truly social* which establish communities, each individual contributing a portion of labour *pro bono publico.* Colonies of bees, of wasps, of termites, of weaver birds, of beavers, etc., afford us examples of what we here term social animals, for they labour in a common cause. Wild cattle, deer, and antelopes, etc., are gregarious; they herd together without any definite purpose, and yet form troops, each individual of which is acknowledged as a member of the community. Shoals of fishes are, in our acceptation of the word, congregational. Guided by instinctive impulse, they associate together in millions, but they act not in concert as do rooks or wild fowl, but even devour each other; none in the advancing phalanx gives the signals of danger to the succeeding hordes. Most of our migratory birds are truly gregarious only for a season; bound together by a voice which says it is time to collect and depart, they obey the call, and wing their way in assembled thousands. We may here adduce the lemming of northern Europe, which occasionally, in myriads, migrating from its native region, devastates the cultivated country. But no tie binds these animals to each other. They unite together only on that instinctive impulse which causes the swallows to assemble ere they take their flight. Among birds of prey, some, as certain vultures, are said to be sociable; they are merely gregarious; so also is the wild rock dove of the cavern. The subject thus briefly touched is not often noticed by Zoologists.

In all our researches into the phenomena of the vital principle we must never forget its Omnipotent Creator. By whatever secondary causes he may work, God alone is the Author and Giver of life. At his fiat, his creatures live; at his fiat, they die. Immeasurably and inconceivably glorious must He be at whose bidding results so wonderful proceed.

There is, however, one species of life higher than any which these pages have been considering—a life, indeed, without which the most exalted form of intellectual existence is comparatively poor and worthless. To understand this subject, however, we must refer to the pages of the word of God. From these unerring records we shall learn that man is by nature dead in trespasses and sins; that created originally in the image of God, with faculties adapted to love his Creator, and to observe those laws which he appointed for his creatures' happiness, man has by transgression fallen from his high estate, and is now with a corrupted nature alienated from God, "an enemy to him by wicked works."

But the Divine record has not been confined to an explanation of the state of spiritual death into which man has fallen. It has pointed out also the remedy which supreme Wisdom has provided. For the violation of God's holy law a full satisfaction and atonement has been made by the death of the Just for the unjust—by the sacrifice of the only begotten Son of God. For the restoration of the soul to happiness and holiness, an effectual provision has also been made in the regenerating and sanctifying influences of the Holy Spirit. These are all essential. Before a man can enter the kingdom of God, he must be born again. A new spiritual life must be breathed into his soul. No form, no sacraments, can by any inherent efficacy accomplish this work. It is the prerogative of the Holy Spirit alone. But this blessing God is willing freely and abundantly to impart to all who, truly feeling their want of it, seek it in his own appointed way, namely, by believing in Jesus Christ, the Son of God, the Saviour of the world. "Let him that is athirst come. And whosoever will, let him take the water of life freely." May it be the reader's happy lot to close with this gracious invitation! As a lost but penitent sinner, in humble faith in the Saviour, yielding himself up unreservedly to him, and seeking, in fervent prayer, the purifying and sanctifying influences of the Holy Spirit, may he be made partaker of that Divine life, which brings with it happiness here and the glories of eternity hereafter!

www.ingramcontent.com/pod-product-compliance
Lightning Source LLC
LaVergne TN
LVHW011224110826
845150LV00006B/1540

9781425515751